Google Earth: Outreach and Activism

Google Earth: Outreach and Activism

Catherine Summerhayes

Bloomsbury Academic
An imprint of Bloomsbury Publishing Inc

B L O O M S B U R Y

NEW YORK · LONDON · OXFORD · NEW DELHI · SYDNEY

Bloomsbury Academic

An imprint of Bloomsbury Publishing Inc

1385 Broadway	50 Bedford Square
New York	London
NY 10018	WC1B 3DP
USA	UK

www.bloomsbury.com

BLOOMSBURY and the Diana logo are trademarks of Bloomsbury Publishing Plc

First published 2015
Paperback edition first published 2016

Library of Congress Cataloging-in-Publication Data
Summerhayes, Catherine (College teacher)
Google earth, outreach and activism / Catherine Summerhayes.
pages cm
Includes bibliographical references and index.
ISBN 978-1-4411-3979-5 (hardback : alk. paper) 1. Geographic information systems.
2. Geographic information systems–Social aspects. 3. Google Earth. I. Title.
G70.212.S86 2015
322.40285–dc23
2014034750

ISBN: HB: 978-1-4411-3979-5
PB: 978-1-5013-2002-6
ePub: 978-1-4411-3497-4
ePDF: 978-1-4411-4780-6

Typeset by Integra Software Services Pvt. Ltd

Contents

Acknowledgements

I wish to acknowledge the following people who have edited or read and commented on or even just talked with me about my writing the material in this book. They have provided very necessary intellectual company and encouragement, I thank them very much: Michael Edmunds and Stuart J. Murray from *Media Tropes*; Hart Cohen and Rachel Morley from *Global Media Journal Australian Edition*; my colleagues Craig Hight, Kate Nash and Fiona Jenkins; Nandita Sharma, my research assistant in the early stages of writing the book (as funded by the College of Arts and Social Sciences, Australian National University); David Dumaresq; Rebecca Moore from Google Earth Outreach; and Katie Gallof and Mary Al-Sayed from Bloomsbury Publishing. I also wish to acknowledge my colleagues who regularly attend the Visible Evidence Documentary Studies Conference. Over the last five years, this conference has been a wonderful forum for presenting and gaining feedback on my research into Google Earth.

Sections in this book can also be attributed to articles I have previously published on Google Earth. Chapters 1 and 4 contain material from my essay 'Google Earth and the Business of Compassion', published in *Global Media Journal: Australian* 4, no. 2 (2010): 1–14. Chapters 4 and 6 draw on 'Embodied Space in Google Earth: *CRISIS IN DARFUR*', *Media Tropes* 3, no. 1 (2011): 113–134. Chapter 5 extends my paper 'Web-Weaving: The Affective Movement of Documentary Imaging', published in *New Documentary Ecologies. Emerging Platforms, Practices and Discourses*, eds Kate Nash, Craig Hight and Catherine Summerhayes (London: Palgrave Macmillan, 2014).

1

An Introduction

Gulliver or Alice would say: 'I've been changed several times.' But for Alice the visible world does not run up against the screen of the mirror; the luminous reflective is not a limit but a point of passage.

Paul Virilio[1]

One of the major challenges that society faces at this current historical moment is to understand how we embody our perception of the world via digital technologies. Actual people and places populate this world that is represented to us as existing in a new kind of communicative space. We are faced with the question of how we are developing a new kind of social intelligence that incorporates the information we gain through the mediated yet always 'real' spaces available to us via digital media platforms that are created from corporately owned software networks and systems. Many of these software systems have till now only been minimally critiqued for their social implications. The first tenet that underlies the argument in this book is that there is an imperative to understand how social networking and information websites work in socio- and geopolitical contexts, if society is to use these sites effectively and for the public good.[2] The second tenet is that Google Earth can be considered one of these massively social media platforms. The third tenet is one that has been and continues to be debated

---

[1] Paul Virilio, *War and Cinema. The Logistics of Perception*, trans. Patrick Camiller (London: Verso, 1989), 32.

[2] My use of this term draws on a simple definition of a term that is now derived from economic theory: '(a) the welfare of the community as a whole, public interest; (b) a commodity held in common (usu. in pl.); (Econ.) a commodity or service provided, without profit, to all members of a society (whether by the government or privately)'. *Oxford English Dictionary* (Oxford University Press, 2013) http://www.oed.com, stable.

in both the scholarly and public sphere: that digital technologies have affected human sociality – how we as humans relate to each other, to our communication technologies and to the rest of our environment. In Sherry Turkle's words:

> We are witnessing a new form of sociality in which the isolation of our physical bodies does not indicate our state of connectedness but may be its precondition.[3]

Such a 'new form of sociality' suggests that we must once again deal with questions about meaning and interpretation that arise from collisions between textual representations and the actual worlds they indicate, in other words and once again, the problematic of what is conceptualized as indexicality. Do we still read texts as if they point directly to or are transparent renditions of the actual world? How much can we rely on previous regimes of interpretation as ways into understanding digitally produced images and dialogues? Nowhere are these questions more relevant and perhaps even urgent as when they are introduced into the context of interactive surveillance software platforms such as Google Earth. At stake are the subjective states of real people: the relationships between us (the viewers/users) and those we connect with via digital technologies. How do we, how *can* we now include our acts of communication in these spaces as fields of actual engagement, affect and effect in the world as we experience it?

CONNECTION with (nearly) anyone at any time: Web 2.0

The most significant and easily understood recent stage in digitally enabled communications arrived with the emergence of what became known as Web 2.0. This term is associated with peer-to-peer interactivity and user participation in data creation. It heralded a new era of the web use 'by the people for the people'. Coined by Darcy Dinucci in 1999, Web 2.0 referred to interactive software platforms and their application to create massive

---

[3] Sherry Turkle, 'Tethering', in *Sensorium. Embodied Experience, Technology and Contemporary Art*, ed. Caroline A. Jones (Cambridge, MA: MIT Press, 2006), 222.

interactive communication sites now known as social media websites. The term was popularized by Tim O'Reilly at the O'Reilly Media Conference in 2004. In a paper published in 2007, O'Reilly defined Web 2.0 as a development of the overall architecture of the World Wide Web (WWW):

> Web 2.0 applications are those that make the most of the intrinsic advantages of that platform: delivering software as a continually-updated service that gets better the more people use it, consuming and remixing data from multiple sources, including individual users, while providing their own data and services in a form that allows remixing by others, creating network effects through an architecture of participation, and going beyond the page metaphor of Web 1.0 to deliver rich user experiences.[4]

Principal architect of the WWW in 1991, Tim Berners-Lee took exception to this definition. He maintained that the WWW was always a 'two-way medium', a platform for interconnectivity and that the term was simply jargon and did not signal a 'reorientation' at all.[5] Nevertheless, the term 'Web 2.0' has become a marker to denote massively popular social media sites in the West such as Facebook and YouTube, the Chinese microblogging site Sina Weibo and the Chinese social networking site Renren. We now *expect* to be able to contact anyone anywhere in the world at any time, no matter where we are and no matter the inconvenience of different time zones. We *expect* this kind of connectivity to be easily accessible via our personal computing technologies and that other people will make themselves available for connection with us. And we *expect* this technology to be easy to use, to be seamlessly able to transport and link us via our comments, concerns, questions and desires, to any database that seems useful to us.

Along the continuing cyber highway of Web 2.0, the makers, the coders, the hackers and the watchers in business and government have inevitably created huge databases of information about people, what they like, what they look like and what they have done. Although the popular press at the

---

[4] Tim O'Reilly, 'What Is Web 2.0: Design Patterns and Business Models for the Next Generation of Software', *Communications and Strategies* 65, no. 1 (2007): 17.

[5] José van Dijck, *The Culture of Connectivity. A Critical History of Social Media* (Oxford, NY: Oxford University Press, 2013), 177n5.

time of writing is beginning to talk about 'the right to anonymity', Web 2.0 with its inherent cultural counterpart that van Dijck calls 'the culture of connectivity' nevertheless demands varying levels of surrender of personal privacy. It is very difficult to envision a cultural shift from this current imperative to connect freely and randomly with each other online. An obvious consequence is the use of Web 2.0 for identity theft, fraud and many levels of warfare. The ongoing surge of cybercrime is resulting in a running 'game of tag' between corporations and governments and cyber criminals, with new forms of security and encryption becoming highly prized forms of currency in the domains of business and government. This desperate quest to secure portals to the WWW against subversion and surveillance is not just a by-product of Web 2.0: it is an inherent characteristic. Such characteristics can be seen as flaws in the utopia of peer-to-peer engagement on the WWW, but they can also be understood to be the parameters of usefulness and capacity through which software creators need to push and extend the programs associated with Web 2.0, if this current culture of connectivity is to continue and to evolve.

Although O'Reilly defined Web 2.0 in the context of business models, what we now understand to be Web 2.0 are all those software applications and platforms that allow us to interact with each other online. O'Reilly writes that a company that promotes itself as using Web 2.0 needs to address the following benchmarks:

- services, not packaged software, with cost-effective scalability
- control over unique, hard-to-recreate data sources that get richer as more people use them
- trusting users as co-developers, harnessing collective intelligence
- leveraging the long tail through customer self-service, software above the level of a single device
- lightweight user interfaces, development models and business models[6]

His second and third points are the most interesting insights in relation to social media sites and are highly significant for thinking about and analysing Google Earth and carefully produced interactive databases

---

[6]   O'Reilly, 'What Is Web 2.0', 37.

such as Wikipedia. This book proposes that Google Earth most certainly is a site that is 'harnessing collective intelligence'. Google Earth also demonstrates how difficult it is to maintain 'control over unique, hard-to-recreate data sources that get richer as more people use them' whilst at the same time offering a format for information and dialogue, which is clear and easily accessible to all users. As described in Chapter 2, this challenge has led to some recent developments in the organization of Google Earth layers and its incorporation of Google Maps. With Google Earth, Google has created a giant platform for connecting people and information. Harnessing resulting data to the needs of users and of its own business model requires constant modifications and updates that defy any sense of permanence and stasis.

A numinous virtually real world

Google Earth was launched by google.com in June 2005, towards the end of a digital media culture that commonly described communication undertaken via the web as happening in 'virtual space'. Prior to the emergence and social dominance of Web 2.0 with the introduction of the major social networking sites Facebook and YouTube (also launched in 2005), virtual space held the same kind of numinosity as William Gibson's 'cyberspace'[7] – an alternate reality. This earlier and increasingly irrelevant meaning of the word 'virtual' implied a separate space of existence which was more related to imagined worlds than the live 'meat' worlds of mundane human existence.

Challenging the somewhat wishful thinking by some people even now that there is virtual space which is separate from the rest of existence, new media theorist Anna Munster re-conceptualized the virtual, soon after the introduction of Web 2.0, as a 'set of potential movements produced by forces that differentially work through matter, resulting in the actualization of that matter under local conditions'.[8] The general, and unfortunately, belated public

---

[7] The term 'cyberspace' was first used by William Gibson in his science fiction book: *Neuromancer* (New York: Ace Books, 1984).

[8] Anna Munster, *Materializing New Media. Embodiment in Information Aesthetics* (Hanover, NH: University Press of New England, 2006), 90.

realization has grown over the last five years or so, that virtual spaces not only are imagined to be real but also have real-life consequences. For example, the problem of material posted on Facebook sabotaging relationships and employment opportunities is now well known. Media theorist Sean Cubitt well describes the not so obvious characteristic of cyberspace as a space of possibilities and responsibilities rather than a simplistic vehicle for 'play acting'; he describes how this digitally created space of communication is one of process rather than an encapsulated space with little or no relation to the actual world:

> In this region of cyberspace (which, like deep space, is lumpy), the real has not faded. It has been registered in the raster of the accountable future, not yet an object of knowledge but for which a place of reckoning has been prepared.[9]

It is interesting that while Facebook claims ownership of all images posted on its site,[10] it has now announced that users can take down, 'delete' their site, although it also says that information may still exist on backup files of Facebook.[11] The introduction of this facility to delete your files illustrates

[9]   Sean Cubitt, *Digital Aesthetics* (London: Sage Publications, 1998), 50.

[10] Facebook.com © 2013,
      'Sharing Your Content and Information
      You own all of the content and information you post on Facebook, and you can control how it is shared through your privacy and application settings. In addition:

      1. For content that is covered by intellectual property rights, like photos and videos (IP content), you specifically give us the following permission, subject to your privacy and application settings: you grant us a non-exclusive, transferable, sub-licensable, royalty-free, worldwide license to use any IP content that you post on or in connection with Facebook (IP License). This IP License ends when you delete your IP content or your account unless your content has been shared with others, and they have not deleted it.
      2. When you delete IP content, it is deleted in a manner similar to emptying the recycle bin on a computer. However, you understand that removed content may persist in backup copies for a reasonable period of time (but will not be available to others)'. https://www.facebook.com/legal/terms (accessed 29 September 2013).

[11] Facebook.com © 2013,
      'Deletion
      When you delete an account, it is permanently deleted from Facebook. It typically takes about one month to delete an account, but some information may remain in backup copies and logs for up to 90 days. You should only delete your account if you are sure you never want to reactivate it. You can delete your account here. Learn more.
      Certain information is needed to provide you with services, so we only delete this information after you delete your account. Some of the things you do on Facebook aren't stored in your account, like posting to a group or sending someone a message (where your friend may still have a message you sent, even after you delete your account). That information remains after you delete your account'. https://www.facebook.com/about/privacy/your-info (accessed 29 September 2013).

well the current growing demand for mutability and agency by users of social media. Our many human endeavours in the actual, whether they are individual or group based, social, commercial, activist, altruistic or in the realm of the perceived 'good' of nation states all still depend on how we perceive our actions in what used to be (and often still is) called virtual space or cyberspace: that realm of communicative texts and processes enabled by the Internet and the WWW

Be it a reference to the magical, the terrible, the criminal, altruistic or simple escapism, I propose that the meaning of the word 'virtual' is still relevant however, in our quest to find out more about the ever-shifting *limen* between how we can use and understand more about what we do 'on the web', and how that 'doing' simultaneously occurs in the actual world. Since 2005 the term 'magical' perhaps has become a relevant descriptor for the virtual as people immerse themselves in the WWW via social media platforms (Facebook and YouTube being the dominant players at this current time) and many complex world-creating online games and spaces such as World of Warcraft and Second Life, although, as noted earlier, escapist ideas about abandoning the 'real world' for the virtual still allow an elusive yet obstinate illusion that people can be anyone or anything they like in virtual space, and that actions in this space hold no or few consequences in people's otherwise actual lives. The Social Web, Web 2.0, not only introduced new digital platforms for networking and community formation but also heralded the arrival of a young generation of users who have grown up with these websites and who are becoming more aware of (if not fast enough, even now) the life implications of social media. One of the primary premises of this book is to examine such 'actualization of that matter under local conditions', which Munster says is implied by the use of this term 'virtual', as it comes about in experiencing Google Earth.

Google Earth is a major and highly accessible player in the mapping activities and data visualization that cartographer Jeremy W. Crampton describes as the 'geoweb' or 'spatial media':

distinctly public and citizen orientated mapmaking efforts, which raises interesting questions not only about access and control of the geographic information but of the possibilities for counter-mapping and counter- knowledge.[12]

Google Earth offers Haraway's 'god-trick' – the illusion of infinite vision – to all of us, to download for free (see Figure 1.1 for an image of Google Earth's spinning globe). In 2007, Jo Wood et al. noted the emergence of 'time-line functionality allowing elements with temporal information to be encoded (KMLv2.1) and then selected and filtered by the user'[13] and that as a result, 'data can also be streamed from a server in response to changes in the visible area of the viewing window sent by Google Earth'.[14] A user of Google Earth then can interact with the website via hyperlinks to other sites, and through the appealing simulation of travel that shows the user's journey through the site as flying over and into the oceans and landscapes

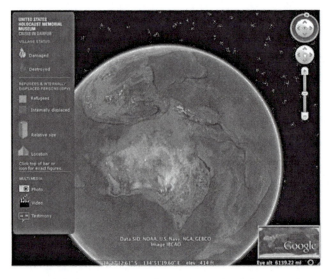

Figure 1.1 Screenshot from opening of Google Earth, Google Earth© captured 18 September 2010. Google and the Google logo are registered trademarks of Google Inc., used with permission.

12 Jeremy W. Crampton, 'Cartography: Maps 2.0', *Progress in Human Geography* 33, no. 2 (2009): 91.
13 Jo Wood, Jason Dykes, Aidan Slingsby and Keith Clarke, 'Interactive Visual Exploration of a Large Spatio-Temporal Dataset: Reflections on a Geovisualization Mashup', *IEEE Transactions on Visualization and Computer Graphics* 13, no. 6 (2007): 1177.
14 Wood et al., 'Interactive Visual Exploration of a Large Spatio-Temporal Dataset', 1177.

of earth as represented by an animation of the beautiful blue globe of planet Earth, as if it were seen from space.

Methodology: Hermeneutic framework and fieldwork

This book belongs within the discourse of Cultural Studies which Donna Haraway defines as follows:

> A set of discourses about the apparatus of bodily/cultural production; emphasis on the irreducible specificity of that apparatus for each entity. Not culture only as symbols and meanings, not comparative culture studies but culture as an account of the agencies, hegemonies, counter-hegemonies, and unexpected possibilities of bodily construction.[15]

The task I have set myself in this book requires a wide theoretical landscape where a discussion can move flexibly between a number of concepts that can be used to illuminate some of the problems inherent to studying a digital object like Google Earth. My methodology primarily lies in the humanities context of textual analysis, based in the self-reflexive tradition of hermeneutics.[16] This methodology uses the act of critical personal interpretation as a way into understanding. In Gadamer's words, a text is 'that which resists integration in experience and represents the return to the supposed given that would then provide a better orientation for understanding'.[17] My interpretation of Google Earth as a new kind of text emphasizes the word 'resists' in this quote. The images derived from Google Earth are the textual sites for interpretation – Gadamer's 'supposed given'. These images shown on Google Earth web pages 'resist' the experience

---

[15] Donna J. Haraway, 'A Game of Cat's Cradle: Science Studies, Feminist Theory, Cultural Studies', in *Critical Digital Studies: A Reader*, Second Edition, eds Arthur Kroker and Marilouise Kroker (Toronto: University of Toronto Press, 2013), 64.

[16] A definition of the adjective 'hermeneutic' from the *Oxford Online English Dictionary* (2013): 'Belonging to or concerned with interpretation; esp. as distinguished from exegesis or practical exposition...' and as a noun: 'a method or theory of interpretation', *Oxford Dictionaries Online*, http://oxforddictionaries.com/definition/english/hermeneutic (accessed 28 September 2013).

[17] Hans-Georg Gadamer, 'Text and Interpretation', in *Hermeneutics and Modern Philosophy*, ed. Brice R. Wachterhauser, trans. Dennis J. Schmidt (Albany, NY: State University of New York Press, 1986), 389.

of immediacy in the sense that they make statements about what is, what exists in the world, as if these statements have the authority of a static knowledgeable source. As with all texts that we interpret, they are part of a dominant hegemonic ideology about how the world is and should be. In other words, when thinking of something as a text, we are engaging with it as an entity in its own right. So a text belongs both in the time we engage with it (the present times of our interpretation) and the times in which it is produced and shown to us. In Chapter 3, I also include in an account of my own specific 'fieldwork' experience in Google Earth.

Virtual objects on the WWW and the Internet (e.g. websites, computer-generated images and emails) are newly amazing to us still in many ways, but perhaps most particularly because they are presenting themselves as texts that depend for their existence on the paradox that is 'time'. We often use our concept of time as a historical construct in order to fix an object for our focused attention. But when we look at websites, the very processes of change and movement through time and place are more evident than ever before. Websites can be changed very easily, and this possibility of change challenges our confidence in being able to exactly 'pin down' the meaning of these digital texts. The challenge is to understand the varying levels of a new kind of realism that affects our perception of the world as it is represented on the WWW and the Internet. The new kind of sociality that Turkle describes depends on our slowly growing ability to accept that digital technologies are offering us new ways to perceive the world. I suggest that these new ways are nevertheless based to a degree on how we as a culture have come to learn our world through the lens of a camera.[18] My analyses in this book draw particularly on noting ways in which Google Earth invites several kinds of embodied reception that relies on the 'lens' of remote sensing satellite technologies.

Websites derived from remote sensing surveillance technologies such as Google Earth even subsume their changing images of visualized time and space into their textual 'given'. Yet at the same time, the sites for interpretation offered by Google Earth demand a new consideration of the ways in which

---

[18] Lev Manovich, *The Language of New Media* (Cambridge, MA: MIT Press, 2000), 244–276.

the illusion of direct experience – that sense of personalized contact in a particular time and space – works. Google Earth draws us into an imagined space of contact with the people and places it represents. The imagined space of contact is nevertheless also embodied via the very mechanisms that provoke our perceptual immersion.

This imagined, conceived space of contact is analogous to the spaces for interpretation offered through our cinematic literacy, a still evolving literacy in audio-vision that began over 100 years ago. The practice of cinema, including its making, presentation and reception, relies on modes of performance which proceed 'through' time rather than 'at the same time'. This idea of performances that move through time rather than just at a particular time recalls Gadamer's description of time as supporting the process of understanding:

> Time is no longer primarily a gulf to be bridged, because it separates, but is actually the supportive ground of process in which the present is rooted. Hence temporal distance is not something that must be overcome.[19]

The concept of time as a 'supportive ground of process' is particularly relevant for interpreting Google Earth: a site which embeds its own authority as a 'truth platform' through an assumed authority of showing how time progresses.

By using a hermeneutic method to understand Google Earth then, this rendition of how time progresses needs also to be considered as itself a text that can be understood as a site for interpretation, as well as all the many and various acts of reception and making that are part of the site. When Google Earth is seen as a new kind of interactive text, the process of hermeneutic analysis is drawn away from looking at a text as something that has been made in the past and considered in the present, towards a process of interpretation that includes a search for ways of *critiquing* text that situates itself as a continuing present.

---

[19] Hans-Georg Gadamer, 'The Historicity of Understanding as Hermeneutic Principle', in *Heidegger and Modern Philosophy*, ed. Michael Murray (New Haven, CT: Yale University Press, 1978), 181.

My critical stance in this book draws not only on the personalized practice of hermeneutic interpretation but on the modes of critical practice to be found in current software studies, affect studies, documentary studies and cinema studies. In Chapters 5 and 6, I use the idea of *gest* from Brecht's theory of performance and theatre, as a way into describing some of the embodied ways we can relate to Google Earth.

My overall approach in writing this book is to provide a close reading of Google Earth as a powerful player in the communication domain of social media. Whilst the whole book addresses this idea, the question is posed most literally in Chapter 2. Each chapter addresses an aspect of the site. First, in Chapter 2, I extend my discussion of Google Earth's aesthetic in the context of its military prehistory, its construction, its links to other similar google-linked world-making sites such as Google Maps and how it is perceived critically by social scientists. In Chapter 3, I overview some of the many ways in which Google Earth is being used: by the general public as a personalized tool for communication and diary-making, by community groups, by the site's official Google Earth Community, by unofficial blogs and by social science researchers.

Chapters 4 and 5 survey the Google Earth Outreach Program and how the site is used by human rights and environmental activist organizations. These uses include showcasing their activities and for more generally using Google Earth as a distribution site for information about significant areas of concern in human rights and environmental discourses. Human rights activism via Google Earth is of particular interest and focus. Lisa Parks in her landmark essay 'Digging into Google Earth: An analysis of "Crisis in Darfur"' (2009) introduces the study of affect as a way into understanding receptive possibilities in experiencing Google Earth:

> few have considered how Google Earth builds upon and differs from earlier global media formats and how it structures geopolitics as a 'domain of affect', particularly when used as a technology of humanitarian intervention.[20]

---

[20] Lisa Parks, 'Digging into Google Earth: An Analysis of "Crisis in Darfur"', *Geoforum* 40, no. 4 (2009): 535–545.

Indeed, her most recent critique (2013)[21] offers perhaps more optimistic insights into how the site can manifest as an opportunity for expanding practices aimed at the social good. Her commentary and Munster's (2013)[22] provide articulate and comprehensive platforms of comparison for my own discussions.

In Chapters 6 and 7, I argue that changes in perception are available to users of Google Earth because of the site's content, its bioconvergent aesthetic and the ways in which its technology allows the user to locate a personal sphere within an enormous, politically charged global environment.

Another kind of vision

The 'eyes' made available in modern technological sciences shatter any idea of passive vision; these prosthetic devices show us that all eyes, including our own organic ones, are active perceptual systems, building in translations and specific *ways* of seeing, that is ways of life …

Donna J. Haraway[23]

In the above quote (and leaving arguments about prosthesis aside), Haraway conceptualizes this kind of active vision as another kind of 'gaze' which might not necessarily involve the same plays of power that have traditionally been accepted as part of 'looking at' people different from ourselves. Instead, these new kinds of eyes might invoke another kind of agency that is not perhaps explicitly political. The interactivity between human and machine that is required in order to 'look at' Google Earth is aggressive in the sense that it is purposeful. It is not possible, however, to clearly differentiate between various motivations that might lie behind a particular instance of accessing Google Earth. These motivations can include an open curiosity or a more focused and sometimes unacknowledged

---

[21] Lisa Parks, 'Earth Observation and Signal Territories: Studying U.S. Broadcast Infrastructure through Historical Network Maps, Google Earth, and Fieldwork', *Canadian Journal of Communication* 38 (2013): 304.

[22] Anna Munster, *An Aesthesia of Networks* (Cambridge, MA: MIT Press, 2013), 45–72.

[23] Donna J. Haraway, *Simians, Cyborgs and Women. The Reinvention of Nature* (London: Free Association Books, 1991), 190.

curiosity aimed at finding certain sites. In turn, curiosity can be derived from and provoke morally positive or negative points of view. Constructive action towards the well-being of another individual can be an outcome of searching Google Earth, but this of course is only one possible outcome amongst many others.

Google Earth is constructed from a culmination of remote sensing satellite technologies, mega database and computing power, and 3D animation. In this book, I maintain that it is both a tool for militarized vision and also a tool for embodied compassionate vision, in artist Caroline Bassett's words, for 'love at a distance'.[24] It is tempting to extend Bassett's claim that this kind of love which can be accessed through remote sensing technologies towards an esoteric reception of Google Earth that is akin to descriptions of the 'wonder' induced oneiric film (a film with dreamlike aesthetics). Such a flamboyant description of the images we find in Google Earth is not totally beside the point. Cubitt's writing on oneiric film recalls, for me at least, some of the wonder I still feel in sensing my vision flying around Google Earth's depiction of our beautiful blue planet:

> The sense of wonder evoked in oneiric cinema is a matter of style as self-derealization – a special moment in the history of wonder … Like every instinct, wonder becomes socialized as a drive.[25]

Cubitt describes wonder as a state in which the subject's sense of self is subsumed in a sensation of apprehending, and he distinguishes it from the sublime as follows:

> Wonder is entirely bathed in time, in the instant of 'before'. The sublime stands above and beyond time, proposing good and evil as equally ineffable and absolute binarism enacted in the television images of the September 11 events. Wonder on the other hand, is dialectical …[26]

My discussions in this book tend towards a description of Google Earth that aligns the site with the affect of wonder, rather than the more static,

[24] Caroline Bassett, 'Remote Sensing', in *Sensorium. Embodied Experience, Technology and Contemporary Art*, ed. Caroline A. Jones (Cambridge, MA: MIT Press, 2006), 201.

[25] Sean Cubitt, *The Cinema Effect* (Cambridge, MA: MIT Press), 297.

[26] Cubitt, *The Cinema Effect*, 324.

structured vision of the sublime. It also proposes that the affect of wonder in experiencing Google Earth has the potential to be instrumental for the public good.

A political aesthetic

Politics revolves around what is seen and what can be said about it, around who has the ability to see and the talent to speak, around the properties of spaces and the possibilities of time.

Jacques Rancière[27]

Rancière's above-quoted idea of the relation between politics and 'the ability to see' as embedded in 'properties of spaces and the possibilities of time' introduces three elements integral to new visions of the world enabled by Google Earth. This 'ability to see' is hegemonic. It relies on how we are taught by society to see and relies on *how* we are shown, as well as what we are allowed to see and who is making that which we are invited to see. As communications scholar Roger Stahl notes:

the geopolitical significance of *Google Earth* can be approached from at least two directions... [as] a 'metaregime of visibility...' with the second perspective operating via 'the aesthetics of visibility' or the ways *Google Earth* acts as a kind of text, powerful public screen onto which a political landscape is projected and thereby made sensible.[28]

Whilst Chapters 2 and 3 inform on the histories, corporate strategies and common uses of Google Earth, my discussions in Chapters 4 and 5 specifically address Stahl's latter perspective in the contexts of aesthetics, affect and ethics. In these chapters, I particularly argue that one of the responses available to texts made available via Google Earth can be named, in philosophical terms, as the affective emotion of compassion. I draw here on Martha Nussbaum's

---

[27] Jacques Rancière, 'The Distribution of the Sensible: Politics and Aesthetics', in *The Politics of Aesthetics*, trans. Gabriel (Rockhill, SC: Continuum, 2004), 13.

[28] Roger Stahl, 'Becoming Bombs: 3D Animated Satellite Imagery and the Weaponization of the Civic Eye', *Mediatropes* 2, no. 2 (2010), 67. ISSN 1913–6005.

understanding of compassion as a 'thought experiment' and as a 'goodwill' towards others' well-being. Compassion, then, is an emotion-driven imagination that draws on a person's own life experience to identify with and, to an extent, feel another's situation.[29] I propose that such an approach is valid because it adds to our knowledge of how the 'political landscape' of such a widely accessible text as *Google Earth* is understood by people using it as an individualized point of access to the actual world.

Stahl's 2010 essay and my approach share a joint interest in 'how the Google Earth aesthetic has evolved to become a part of public consciousness'.[30] The playful delight of simultaneously flying and looking and yet also finding out serious information is easily available in the experience of Google Earth; aesthetics become not only the domain of institutionalized, militarized reportage but also the domain of purposeful individual imagination and understanding. The concrete, haptic interactivity involved in accessing Google Earth via a personal computer is clearly the primary trigger for such immersions that also depend, as do all receptive experiences, on our cognitive interactions with images – on the connections we make between images and our personal histories and memories. I propose that compassion is one possible response to some web pages in Google Earth and that compassion constitutes a particular kind of our 'ability to see', one of the several subjective processes whereby we understand the complex textual instances found in Google Earth. My case study analyses will draw particularly on noting ways in which Google Earth invites several kinds of embodied reception.

The variables in our physical acts of searching are several, including the chronology of how we follow links, where we begin, where we end our browsing and what level of viewing we use via the zoom and tilt tools. Our interpretative acts of vision through searching encompass all of the above but include many more: our prior knowledge of the situation described by the website's content, our digital literacy, our patience, our curiosity, our fascination, how and why we begin and end our search, whether or not we return and what form that

---

[29] Martha Nussbaum, 'Compassion: The Basic Social Emotion', in *The Communitarian Challenge to Liberalism*, eds Ellen Frankel, Fred D. Miller Jr. and Jeffrey Paul (Cambridge: Cambridge University Press, 1996), 28, 48.

[30] Stahl, 'Becoming Bombs', 67.

return takes. The current world of interactive digital environments demands from us a high level of digital literacy if we are to use them in order to exert any kind of political agency. Such literacy includes the understanding of the 'virtual flyby' sensation which developed from 'first-person shooter' games that still offer a banal yet enticing sense of power, freedom and visceral delight, and in games that combine live action sequences, complex and interactive storylines, Stahl well describes Google Earth in the context of its history and the resulting geopolitics of militaristic 'virtual flyby' aesthetics that are inherent in 'first-person shooter'-styled games. And he focuses his investigation on how Google Earth acts 'as a kind of text, a powerful public screen onto which a political landscape is projected and thereby made sensible'.[31] Stahl is concerned with the tension between human agency and geopolitical forces, with what narratives are recalled in the image, 'and how weaving the technology into existent cultural practices plays a part in conditioning the meaning of geopolitical relations ...'.[32]

A case study

In 2007, Google Earth introduced Google Outreach. The first foray into this plan for outreach came in the shape of Global Awareness Layers created from Keyhole Markup Language (KML)[33] – overlays/sets of tags and links that provide information about a specific issue and sourced from links to and information from activist and not-for-profit organizations. The knowledge we gain from looking at human rights and animal rights activist sites also can include our responses to what Rancière names 'the intolerable image'. He asks 'What makes an image intolerable?'[34] His reply is in the shape of two

---

[31] Stahl, 'Becoming Bombs', 67.

[32] Stahl, 'Becoming Bombs', 67.

[33] Google.com,
'What is KML?
KML is a file format used to display geographic data in an Earth browser, such as Google Earth, Google Maps, and Google Maps for mobile. You can create KML files to pinpoint locations, add image overlays, and expose rich data in new ways. KML is an international standard maintained by the Open Geospatial Consortium, Inc. (OGC)'. https://developers.google.com/kml/ (accessed 1 October 2013).

[34] Jacques Rancière, *The Emancipated Spectator*, trans. Gregory Elliott (London: Verso, 2009), 83.

more questions: 'what features make us unable to view an image without experiencing pain or indignation' and then 'is it acceptable to make such images and exhibit them to others?'[35] The textual analyses that are included in this book incorporate such issues of power, the ability to see and be seen, as well as the overall ethics involved in activist sites that become visible in Google Earth as a Global Awareness Layer.

In Chapters 4 and 5, I introduce my case studies that are drawn from one of the earliest of these layers: a subtext that consists of Google Earth's Global Awareness Layer called 'Crisis in Darfur'. This layer is not a place for leisurely respite of any kind but one which presents the horrors of a recent genocide. Through focusing on Crisis in Darfur, I more closely examine the question: does the way we can interact with Google Earth offer a new pathway for compassion? I suggest that the Google Earth layer Crisis in Darfur, constructed in collaboration with the United States Holocaust Memorial Museum in 2007, offers an opportunity to appraise a new conjunction between those who see and those who are seen.

In Google Earth, both the familiar and the unfamiliar are contained within the same database of remotely sensed images: the far and the near are conflated within the images we find. As noted earlier in this Introduction, I propose that there is a new way of understanding how the act of finding pictures and stories of people who are far away and unfamiliar to us might evoke a feeling of empathy from a user, especially from one who has recently also found within the same website more domestic pictures and information about parts of the earth that are familiar and easily imagined. Both the familiar and the unfamiliar are accessed via the same interactive actions that are required to gain information from Google Earth.

Towards an ethical embodied vision

Issues of power and vision are integral to considering the images and stories of trauma that are made available via Google Earth, and in Haraway's words again:

---

[35] Rancière, *The Emancipated Spectator*, 83.

> Vision is *always* a question of power to see – and perhaps of the violence implicit in our visualizing practices. With whose blood were my eyes crafted?[36]

Haraway concisely refers here to the ethical dilemmas of vision, of 'looking at' people. Even the word 'vision' is laden with all the previous debates in academe about voyeurism: of enjoying the spectacles we look at without reference to their sociopolitical consequences, of looking at people who cannot look back, of the power of looking when others cannot and of finding out information unavailable to others. But there are also other, constructive acts of vision that include ways of 'looking at' via acts of imagining and reflection, and the kinds of looking that are not so readily available using only our 'naked eyes'. They include searching, 'looking for' something, for someone, using whatever tool is available – by combining our eyes and ears with the microscope, telescope, the algorithms of mathematics and digital modelling technologies, microphones, earphones – augmenting and changing our 'naked eyes' to include any machine that helps extend our vision towards a more expansive experience of self and the agendas of that self, a more expansive, ethically based subjectivity. Google Earth, a complex site of production and reception, offers a new kind of social text for critical interpretation.

---

[36] Haraway, *Simians, Cyborgs and Women*, 192.

2

'What Is Google Earth?'

*Computers will talk to anyone, but only the wealthy teach them to speak, to
define what perception might be and what is interesting.*

Sean Cubitt[1]

Cubitt's words in the above quote astutely, if somewhat simplistically,
reduce the moral problems inherent to 'Earth-observing media' to their
basic parameters: the power differentials involved in financial investment,
corporate and/or government ownership of digital computer hardware,
software and programming as well as the consequences of potential if
not actual changes in human perception that digital technologies can
offer. In McLuhan's famous if misquoted words, 'the medium is the
massage', the immersive effects of some digital technologies most certainly
recall the implications of their *massaging* of any embedded message/
agenda.[2] The conflation of media with the content it communicated is one
way to articulate how people can be confused by the spectacular nature of
a communication medium and led into believing or taking at face value
what informational experience that medium can actually mediate. The term
massage also well connotes that process of immersive mediation – the
process of a masked/disguised/non-transparent messaging that is the task
of social scientists and public commentators to critique.

---

[1] Cubitt, *Digital Aesthetics*, 47.
[2] Marshall McLuhan, *Understanding Media, The Extensions of Man* (New York, NY: McGraw Hill,
    1964), 7. In 1967, McLuhan also punned on this famous phrase with the phrase 'the medium is the
    massage'. See Marshall McLuhan, Quentin Fiore and coordinated by Jerome Agel, *The Medium Is
    the Massage. An Inventory of Effects* (Berkeley, CA: Bantam Books/Random House, 1967) Ginko
    Press, 2000.

My own critique of Google Earth, although primarily directed at the textual nature of a web platform, is broadly focused on power relations, on the hegemony that a communication medium operates within. The outcome of such critique usually oscillates between descriptions of a medium as disastrous if not criminal and utopian assessments of how the medium will enable all humanity to prosper, how it contributes to the general social good. With the advent of Web 2.0 circa 2005, people who had access to digital media tended to adopt the more utopian approach. As noted in Chapter 1, the general public has recently begun to take on the less than salubrious outcomes of digital social media. As could be expected of strong scholarship, social researchers such as Sherry Turkle and Lisa Parks have been open to considering both the good and the bad of this new media. While Turkle proposes that a 'new form of sociality' is emerging due to digital technologies of communication, Parks focuses on the need for research into what she names

> *infrastructural re-socialization*: a technological literacy project that urges publics to notice, document, and ask questions about infrastructure sites and become involved in discussions and deliberations about their funding, design, installation, operation, and use.[3]

Both Turkle's and Parks' projects are as necessary to society as they are related to each other. Parks' 're-socialization' project provides an extension of both the ideas and practices underlying Turkle's 'new form of sociality'. Both projects can also be then understood as contributing to a positive, carefully optimistic construction of the new networks of social relations that are becoming possible as a result of digital technologies. Such projects cannot be trivialized as 'wishful thinking' because both are grounded, although in different methodological approaches, in empirical research. In her writings (2005 and onwards), Parks especially and explicitly addresses the militaristic surveillance origins of Google Earth; her work recalls Cubitt's following dictum: 'To investigate a medium is to analyze and synthesize the historical

---

[3] Lisa Parks, 'Earth Observation and Signal Territories: Studying U.S. Broadcast Infrastructure through Historical Network Maps, Google Earth, and Fieldwork', *Canadian Journal of Communication* 38 (2013): 303.

nature of the material mediations that characterize a period in time.'[4] What follows here in this chapter is primarily a discussion of the 'material mediations' of Google Earth, of how it began and continues to be a site for cultural meaning and interpretation, one that is easily accessible (usually!) to anyone using a computer linked to the World Wide Web (WWW).

My discussion of Google Earth in this chapter draws on its prehistory, history, construction, aesthetic and description of the capabilities of the web pages that make up the site.

I do look at the technical specificities of Google Earth, but only briefly. Google.com has published much of this kind of information on their 'About' website, as have various blogs and other websites. My primary focus, however, is on how Google Earth is presented as a digital object: its aesthetic, its makers' presumption of its accessibility and its cultural context both now and at the time of its inception. What is interesting about this site is the company it keeps and has kept since its arrival on the web. In Chapter 5, I will be looking further into its web ecology in the context of human rights activism. To interrogate how the site's aesthetic, its presentation on the web, its tools of representation – the site's continually evolving and elaborate animated modelling and simulation programmes – work in the context of what else is happening on the web, it is necessary to draw to both lesser and greater extents from relevant areas of study: cartography, network theory, communication and software studies. So the latter part of this chapter includes a discussion of how issues raised via these disciplines and areas of study can add to a description of Google Earth as a site that embraces a much wider kind of cultural performance than that usually imagined by the everyday user.

So, *what* exactly is Google Earth?

In the Introduction to The Geological Society of America's (GSA) Special Paper 492 (2012), we read that Google Earth is 'a computer program

4   Cubitt, *The Cinema Effect*, 2.

that integrates a global digital elevation model (DEM) with base surface imagery to create a 3D, mirror-world representation of the Earth (Bailey, 2010)' and that '[t]echnically speaking, GE is only 2.5D as the model is projected onto a 2D computer screen with the appearance of being 3D'.[5] In 2010, geologist and staff member of OpenTopography[6] Chris Crosby posted in the OpenTopography Community blog that Google was beginning to include Light Detection and Ranging (LiDAR) data files in the 'terrain layer' in Google Maps and asked 'when will Google tackle the integration of high-resolution terrain data into the topographic mesh in Google Earth?'[7] Now, in 2014, you can open in Google Earth NCALM[8] LiDAR data as published in Google Earth's 'Earth Gallery' (earth.google. com) in the category of 'Terrain and Elevation', showing 3D topographic visualizations. Google Earth has incorporated laser imaging that can be easily accessed, bypassing the levels of (fairly low) expertise previously required to embed LiDAR data into layers in Google Earth. Google Earth currently publishes and updates more than 20 billion megabytes of satellite and aerial image data.[9]

The GSA Special Paper cited earlier also readily and accurately describes the celebratory nature of Google Earth's important efforts in making available its imagery 'in a timely manner'.[10] As will be seen in the discussion of Google Earth Outreach in Chapter 3, this 'timely manner' includes quick updates of crisis areas. It points out that Google Earth is a site that archives and provides imagery to individual researchers that would be too expensive for them to

---

[5] J.E. Bailey, S.J. Whitmeyer and D.G. De Paor, 'Introduction: The Application of Google Geo Tools to Geoscience Education and Research', in *Google Earth and Virtual Visualizations in Geoscience Education and Research*, eds S.J. Whitmeyer et al. (Boulder, CO: Geological Society of America Special Paper 492, 2012), vii. doi: 10.1130/2012.2492(00).

[6] OpenTopography is 'A Portal to High-Resolution Topography Data and Tools', staffed from the Advanced CyberInfrastructure Development (ACID) group at the San Diego Supercomputer Center at UCSD, UNAVCO and the Active Tectonics, Quantitative Structural Geology and Geomorphology group at Arizona State University. See http://www.opentopography.org/index.php (accessed 17 February 2014).

[7] Chris Crosby, 'LiDAR Beginning to Appear in Google Maps Terrain Layer', in OpenTopography Community Blog, 30 July 2010. http://www.opentopography.org/index.php/blog/detail/lidar_ beginning_to_appear_in_google_maps_terrain_layer (accessed 17 February 2014).

[8] NCALM is the LiDAR Data Distribution Center for the (US) National Center for Airborne Laser Mapping.

[9] Google Earth Outreach Executive (GEOE) Interview, 14 February 2014.

[10] Google Earth Outreach Executive (GEOE) Interview, 14 February 2014.

put together themselves. The site thereby plays an important social role in providing high-resolution imagery of earth via remote sensing satellites, at very little financial cost to the general public, researchers and not-for-profit organizations. It also 'provides a canvas to which users can add their own geospatial data to create dynamic visualizations using Keyhole Markup Language (KML)'.[11] Brian McClendon, Vice-President and Co-founder of Google Earth, recently described Google Earth as consisting primarily of three tools: data (20 petabytes of imagery), computing power and Google Earth users (combined with Google Maps, 1 billion users per month). McClendon says that Google Earth's most powerful function is as 'platform [where] passionate groups and individuals to add their information on top'.[12] In other words, using the complex and highly expensive technologies that create Google Earth, Google provides a new kind of highly sophisticated, innovative social media.

Google Earth brought to you by Google.com

Google.com makes available via many web pages within sites linked from the home pages of Google.com and Google Earth, the history and continual development of Google Earth and its usage. Between the years I began my research in 2008 and the time of writing this book in 2013–2014, Google Earth has changed. The changes have been in the ways Google and other branded applications can be accessed through the Google Earth site, the various extra ways for modelling and animating information, the expansion of the Google Earth Community, the degradation, updating and sometimes subsequent renewal of some existing layers and the introduction of new ones, the availability of more simply accessible marketing opportunities, the formalization of sponsorship for not-for-profit organizations, and the increased simplicity and availability for downloading KMZ files. Google

---

[11] Google Earth Outreach Executive (GEOE) Interview, 14 February 2014.
[12] Brian McClendon, 'Acceptance Speech on being awarded the UNEP Champions of the Earth Award in New York', 16 September 2013. Accessed via vimeo, http://vimeo.com/79463572 (accessed 17 February 2014).

Earth is now one of many Earth-imaging platforms that draw on data from satellite and aerial imaging. It is still, however, the most accessible platform for the general public, and its images can now be downloaded via apps onto the personal technologies of mobile phones and tablets. With these changes and additions, Google has also increased the amount of information on its own website about itself as a corporation and the development of Google Earth. Google presents huge amounts of information about itself and its subsidiaries.[13] Here is a brief synopsis. The domain name Google.com was registered on 15 September 1997. In 2004, Google acquired Keyhole and its application Earth Viewer. In 2004, it also acquired Where2, a GPS navigation system, and launched Google.org 'dedicated to the idea that technology can help make the world a better place'.[14] In February 2005, Google launched Google Maps and in June 2005:

> We [Google] unveil, a satellite imagery-based mapping service that lets you take a virtual journey to any location in the world. Google Earth has since been downloaded more than 1 billion times.[15]

In October 2006, Google announced its acquisition of YouTube, establishing the company as not only a mammoth search engine and advertising business but also a major player in the representational communication technologies that came to constitute Web 2.0: that array of social websites that since 2005, with the advent of YouTube and Facebook.com, have allowed ordinary people to share their private lives with mass audiences around the globe. In June 2008, Google Map Maker was launched to enable people 'to directly update geographic information in Google Maps and Google Earth'.[16] In noting this launch, Google adds that Google Map Maker is thereby 'helping ensure that the map accurately reflects the world' – a huge claim: interesting particularly in the use of the terms 'accurately' and 'reflects'. Google was embarked on the idealistic

---

[13] Google provides a year-by-year (1995 to present) concise history and vital statistics of how Google came about on its own home page that can be accessed via the link: http://www.google.com.au/about/company/history/ (accessed 30 October 2013).

[14] See https://earthengine.google.org/#intro (accessed 28 September 2014).

[15] http://www.google.com.au/about/company/history/ (accessed 2 November 2013).

[16] http://www.google.com.au/about/company/history/ (accessed 2 November 2013).

cartographic quest of accurately reflecting the world. In May 2013, World. Time.com launched a gallery of time-lapse images, powered by Google.[17]

As already noted in Chapter 1, Google Earth is clearly not a static site – not really a surprising statement! All websites have the potential to undergo continual updating, and Google is very clear about this in relation to all its applications. For example, on the web page 'Understanding Google Earth Images' Google answers the questions about when imagery is collected and displayed as follows:

> [Google Earth] acquires the best imagery available, most of which is approximately one to three years old. The information is collected over time and is not in 'real time'. For example, *it's not possible to see live changes in images*. (my emphasis)

This page further suggests that '*To get the latest imagery updates, check out Follow Your World*.'[18] According to the unofficial but very popular Google Earth Blog, imagery in Google Earth is updated twice a month, and Google tries to update each area at least every three years.[19] Co-founder Brian McClendon notes that Google Earth is presented as an active site, not a static archive.[20]

The Google Earth home web page also links to a Help web page, which in turn links to a page titled 'Learn' that gives a drop-down link to another page called Release Notes. This page (available via three to four web links or more, depending on how you search) provides a clear summary of information on the latest changes (updates and upgrades) to the site and why they were made. For example, on 31 October 2013 at 5.34 p.m. EST, the page https://support.google.com/Earth/?hl=en-topic=2376010 referred to an archive of changes made up to and including Earth version 7.1.1.1871 as well as the issues that these changes address. Such information pages as those above, through which Google actually defines and says how it considers the site can

---

[17] See http://world.time.com/timelapse2 (accessed 1 November 2013).

[18] See 'Understanding Google Earth imagery', https://support.google.com/Earth/answer/176147?hl=hu (accessed 8 November 2013).

[19] Google Earth Blog, 'How often does Google update the imagery in Google Earth?' (posted 4 October 2010), http://www.gearthblog.com/blog/archives/2010/10/how_often_does_google_update_the_im.html (accessed 16 February 2014).

[20] McClendon, 'UNEP Award Speech'.

be used, can be thought of as showing the 'bones' of the site. They are well-padded bones, padded with details and data structures, but they nevertheless do give access to how and why Google Earth continues to evolve.

The constant updating of Google Earth over the years since its launch, however, is presented on the site itself as an 'expansion of knowledge', a provision of 'extra information' and IT problem solving, and therefore as an extension of the sites' capability for 'truth saying'. But this process of updating and upgrading is not simply or even only a matter of re-organization or addition of new information. The kind of knowledge that is being extended or expanded when the site is updated and/or upgraded is made available via an offering of 'newer' ways for *perceiving and extracting information* from the data presented on the site. Various upgrades (as opposed to 'updates') also mean that newer versions require us to download the latest version to have access to the site's increasingly spectacular imagery and modes of image making. When we talk about 'versions' of a website or web page, we need to cite a URL and an access date if we want to make a reference to information that might be later updated or disappear (hopefully into a web archive). This needs to acknowledge the constantly changing status of information and communicative spaces available on the WWW, well reflecting the now rapidly shifting fashions in designing and using such spaces of information as Google Earth. It also means that although Google Earth presents to us as a coherent and authoritative database of geospatial information, it is clearly a 'patchwork' of information that is current to Google Earth at the time of our viewing.

The frustration that some people feel because they can no longer see what they could at a previous viewing or because images change in other ways often arises because we are distracted by the coherent information template of an animated globe. It is easy to forget that these animations simulate knowledge that is in a constant state of production. Indeed the only constancy is that there is no stasis in the flow of information and we need to remind ourselves that any perceived 'unreliability' of image via Google Earth actually can be understood to introduce a glimmer of transparency to how the information was collected and algorithmically converted to a visualization of Earth.

Consequently, we also need to update constantly our levels of literacy in comprehending these communicative digital objects in cyberspace, along with the information and potential knowledge that are embedded within them: because we need to have some idea about what it is we are looking at on a site like Google Earth in order to extract its full potential as a site for information gathering. We also can decide to become aware of who is using the platform and how. In Chapter 3 for example, my discussion focuses on Google Outreach as one of the ways through which we can decide to participate in the socialized network ecology that has Google Earth as its hub. Munster makes the point that the social media aspect of Google Earth in fact mostly takes place outside of the image platform, that the social, networked aspect takes place in forums inside the Google Community section of the site and in topics within the Official Google Blog.[21] In her words, 'The "population" of Google Earth in fact chooses to reside elsewhere, *adjacent* to the visual virtual globe.'[22] Although these Google-endorsed forums and other unofficial forums (including YouTube channels)[23] are 'outside' the image space, they are certainly part of the whole Google Earth experience for many people. According to media scholar Mathieu O'Neill, 'Today, direct communication without professional mediation is what defines the "blogosphere".'[24] The easy convergence or conflation of Google's own 'professional' sponsored 'blogs' with unofficial ones is itself interesting. It reflects corporate intention to not only somehow monitor online discussion but also encourage people using the somewhat unpredictable 'blogosphere' to join with the corporation in promoting itself as a worthwhile enterprise that is a necessary part of current society – an unspoken proposition that as citizens of the world, somehow or other *we need to engage with Google Earth*.

The aesthetics of this experience then include visualizations of the Earth produced for the individual travelling over and through the virtual globe, and

---

[21] Official Google Blog, http://googleblog.blogspot.com.au/. See the following topic 'A Picture of Earth Through Time' as an example of a topic on this blog: http://googleblog.blogspot.com.au/2013/05/a-picture-of-Earth-through-time.html (accessed 14 November 2013).

[22] Munster, *An Aesthesia of Networks*, 52.

[23] For example Google Earth Hacks, www.gearthblog.com and YouTube channels including the often weird and wonderful sightings cited in 'Secret Places in' and 'Strange Places'.

[24] Mathieu O'Neil, *Cyberchiefs. Autonomy and Authority in Online Tribes* (London: Pluto Press, 2009), 17.

a socially networked acknowledgement that we are aware subjects of the site itself. As subjects and participants, we are all contained within the world created by this virtual globe, what Munster calls 'an aesthesia of networking, lying with new forms for collective practice, formation, and enunciation'.[25] This complex aesthesia is available to the non-coder individual financially for free (apart of course from the required expenditure on owning or hiring computer hardware and Internet access). And the first aesthetic we meet when looking for Google Earth on the web is its home web page:

Figure 2.1 Screenshot of the home page of Google Earth ©, http://www.google.com/Earth/ captured 5.06 p.m. EST, 30 October 2013. Google and the Google logo are registered trademarks of Google Inc., used with permission.

In Figure 2.1 beneath the simple colloquial English phrases 'Hello, Earth' and 'Stay in touch', we glimpse some of the complexity of Google Earth – these two simple turns of phrase are used to introduce a massive amount of information. This web page at http://www.google.com/Earth/ (31 October 2013) cycles through various quite spectacular satellite-derived images of Earth, including those titled Black Rock City, Nevada via DigitalGlobe, Australia from Space via LandSat, Whitsunday Island, Australia via Aerometrex and the image in Figure 2.1, Tahaa and Raiatea, French

[25] Munster, *An Aesthesia of Networks*, 53.

Polynesia via DigitalGlobe. These images do not only provide elegant and beautiful backgrounds to the web page, they also are there to entice the viewer into the fantastic and somewhat illusionary world of Google Earth imagery. They contribute towards a web page aesthetic that is highly geared towards a combinatory form of advertisement that invites viewers to join in the pleasurable experience of looking at beautiful and beautifully rendered photographs.[26] These images not only advertise Google Earth and more broadly Google.com, they also transparently reference their sources via copyright attributions to particular satellite industries, including those cited earlier. The website also provides links for downloading the basic stand-alone version of Google Earth and Google Earth Pro – the version designed for businesses, with added tools for mapping and analysing data. You need to pay for a license to access Google Earth Pro – and I further address this set of capabilities in Chapter 3. Here, I want to move on from briefly noting the functionalities available within the web interface, to a closer examination of how information content and the discourse created by content can constitute a particular performative/interactive aesthetic.

Moving maps

In her essay 'Welcome to Google Earth: Networks, World Making, and Collective Experience',[27] Munster draws a clear relation between the aesthetic of Google Earth and its ideological, 'world-making' capacities. Whereas Crampton (2009), Chris Russill (2013) and Jason Farman (2010) contextualize Google Earth in relation to the discipline of cartography, Munster bases her discussion on the network capabilities that it embeds within its 'network ecology'. Both network theory and cartography clearly deal with the digital world-making enterprise of Google Earth. My commentary delineates between these different approaches only as a way into clarifying the ways

---

[26] I will not be entering into a discussion *per se* of how Google *markets* Google Earth; this context, although tangentially related, is one that belongs to a discipline that is beyond the reach of my analyses in this current book. I acknowledge however that the aesthetic of this web page is highly geared towards a combinatory form of advertisement via an invitation to viewers to experience.

[27] Munster, *An Aesthesia of Networks*, 45–71.

in which Google Earth has been so far considered as a particular receptive experience and its overall performance in a cultural context. My approach in this section is in the context of cartography, while my case study analyses in Chapters 4 and 5 draw more on concepts of performance and network ecologies.

Earth-orbiting satellites, the 'eyes in the sky',[28] have names and are owned by someone even if, in Parks' and Schwoch's words, details of these ownerships and operations 'from financing to launch to service applications, remain hazy at best'.[29] These writers go on to state that the ordinary person's 'perception of various orbits – even the awareness that there are many possible orbital configurations – is often dim'.[30] This being so, then clearly there is a serious knowledge gap between the makers and owners of satellite technologies and the general public; and as with most knowledge gaps, there is also a related gap in power relations that exist between these two broad groups in society. In 2011, fifty-one countries owned their own satellites, with eleven countries having the ability to launch.[31] Again quoting Parks and Schwoch:

> While satellite technologies are implicated in the production of contemporary global imaginaries, they have historically been developed and controlled by a relatively small number of nation-states and corporate entities, making the technology's associations with the 'global' tenuous, if not specious.[32]

Google Earth delivers spatial information primarily drawn from data collected by remote sensing satellites. In other words, it gives us maps: digital maps from web-based geographical information systems (GIS) that themselves have been available only for a very short time. GIS can make information available for mapmaking via vector and/or raster imaging

---

[28] For an illuminating and de-mythologizing of satellite technology and its transformation into 'space junk', see Lisa Parks essay 'When Satellites Fall: On the Trails of Cosmos 954 and USA 193', in *Down to Earth*, eds Lisa Parks and James Schwoch (New Brunswick: Rutgers University Press, 2012), 221–237.

[29] Parks and Schwoch, *Down to Earth*, 3.

[30] Parks and Schwoch, *Down to Earth*, 3.

[31] Parks and Schwoch, *Down to Earth*, 3.

[32] Parks and Schwoch, *Down to Earth*, 3.

processes. Raster images are constructed of pixels, and vector images rely on images created via mathematical formulation of where images are spatially in relation to previously set up vector points. Digital photography, and hence digital mapping, is based on a raster process.[33]

In the 1980s and early 1990s, collaborations between scientists in computer cartography (available from the 1950s),[34] spatial statistics and analysis, and computer science developed into conversations between people who sought to actually define the concepts and practice of GIS, including their accessibilities and effects on various groups in society.[35] The specific quest to visualize Earth as a whole 3D object, however, began with the technology of analogue photography.

Chris Russill's essay 'Earth-Observing Media',[36] written for the edition of the *Canadian Journal of Communication* issue of the same name, begins with his summary account of whole Earth sensing and photography. He quotes the 1966 question that was delivered to Marshall McLuhan by the 'northern Californian hippy' Stewart Brand[37] via a pin-on button that showed this question: 'Why haven't we seen a photograph of the whole Earth yet?'[38] Russill goes on to describe Brand's utopian mission 'to liberate a single photograph from the bureaucratic confines of NASA'.[39] Brand's aim was utopian because he thought that if people could see planet Earth as a whole, then humans would stop warring with themselves, other creatures and the rest of the 'natural' environment, thereby eradicating most of the world's social and environmental problems. In Russill's words, 'A human situated

---

[33] For a simple and clear description of and distinctions between raster and vector imaging in graphic design, see entries to the Graphic Design Forum topic 'An explanation of Raster vs Vector', http://www.graphicdesignforum.com/forum/forum/graphic-design/resources/89-an-explanation-of-raster-vs-vector (accessed 8 August 2014).

[34] Jason Farman, 'Mapping the Digital Empire: And the Process of Postmodern Cartography', *New Media and Society* 12, no. 6 (2010): 870. doi: 10.1177/1461444809350900, http://nms.sagepub.com/content/12/6/869.

[35] For a comprehensive look into the history, practices and concepts of GIS, see Timothy L. Nyerges, Helen Couclelis and Robert McMaster, eds *The Sage Handbook of GIS and Society* (London: Sage Publications, 2011), doi: http://dx.doi.org.virtual.anu.edu.au/10.4135/9781446201046.

[36] Chris Russill, 'Earth-Observing Media', *Canadian Journal of Communication* 38, no. 3 (2013): 277–284.

[37] See Stewart Brand, 'Why Haven't We Seen the Whole Earth?' *The Sixties: The Decade Remembered Now, by the People Who Lived It Then* (New York, NY: Rollingstone Press, 1977), 168–170.

[38] Marshall McLuhan in Russill, 'Earth-Observing Media', 277.

[39] Marshall McLuhan in Russill, 'Earth-Observing Media', 277.

in space could acquire the correct perception of the Earth, and this was experienced as an epiphany reorganizing consciousness.'[40] Brand's campaign to obtain images of Earth succeeded within months, with images still being contributed into the public domain from many sources, including NASA's continuing updates of its iconic and sublime composite image: 'The Blue Marble'.[41]

Beyond military agendas then, what power relations are implicated in the depiction of the Earth as a 3D globe as if it was seen from outer space? Philosopher Jean Baudrillard articulates a description of power that introduces a more dystopian approach to everyman's desire to visualize Earth as a whole. He says that 'Power itself has for a long time produced nothing but the signs of its resemblance,'[42] and goes on to say that with the disappearance of power itself, people crave and demand '*signs* of power – a holy union that is reconstructed around its disappearance'.[43] While an interpretation of Google Earth as a sign of power is undoubtedly appropriate, I suggest that it is not *only* a sign of power (for and of the people). In a poststructuralist fashion, we can also interpret Google Earth to be a conduit for power that flows back and forth between people and the institutions (and/or the other kinds of groups in society to which they belong) through which they act out their lives. Indeed Google Earth, because of its accessibility, is a fine case study for examining how advantage and disadvantage happen via a global, digital platform for communication. Clearly Google Earth, as does every map, plays into the mapping of new digital empires and as users we are drawn into dealing with new forms of sovereignty and power. Google Earth can itself thereby be understood as a complex cultural text.

The interactive interfaces created in Google Earth are between its web pages (via their existence as culturally meaningful interactive texts) and us – the people who interrogate these texts as we explore them for information and for the pleasurable affect of this exploration itself. In the context of digital

---

40 Russill, 'Earth-Observing Media', 278.

41 See NASA (United States National Aeronautics and Space Administration), http://www.nasa.gov/
multimedia/imagegallery/imaGoogleEarth_feature_2159.html (accessed 6 November 2013). 'The
Blue Marble' was created from images captured by the satellite Suomi NPP on 4 January 2012.

42 Jean Baudrillard, *Simulacra and Simulation*, trans. Sheila Faria Glaser (Ann Arbor, MI: The
University of Michigan Press, 2004), 23.

43 Baudrillard, *Simulacra and Simulation*, 23.

technologies that capture and represent 'raw data', digital media scholar Alexander Galloway makes clear the useful and in many ways obvious distinction that exists between what is gathered as 'data' and what that data is shaped/formed into 'information'. He develops a sophisticated set of these on how we can talk about an aesthetic of the digital interface, including the following statement:

> Thus if data open a door into the realm of the empirical and ultimately the ontological (the level of being), information by contrast opens a door into the realm of the aesthetic.[44]

In other words, the data captured by remote sensing satellites has an existence both as format and as content. Humans choose algorithms – mathematical procedures – in order to interpret and represent that data. It is only at this stage of interpretation and representation that we can talk about how we can access that data, because it is then transformed into a pattern (usually visual) of information at a computer interface. And here, at this interface we can choose various styles and designs for representing that information. In designing websites and web pages, we choose an aesthetic whereby we communicate this information. As Galloway notes, raw digital data is never visual of itself and 'any visualization of data requires a contingent leap from the mode of the mathematical to the mode of the visual'.[45] He is speaking here of the Internet (that enabling matrix of networked connections within cyberspace) and the WWW (digital objects in cyberspace). However, it is worth noting that live theatrical performances involving the colourful on-screen display of *live* coding that simultaneously manifests as electronic music[46] do create an exception to what Galloway describes as the law that 'an increase in aesthetic information produces a decline in information aesthetics'.[47] Nevertheless and exceptions aside, his following description of what happens when we look at a computer screen is

---

[44] Alexander R. Galloway, *The Interface Effect* (Cambridge: Polity, 2012), 82.

[45] Galloway, *The Interface Effect*, 82.

[46] See Ben Swift et al., 'Distributed Performance in Live Coding Stuff', in *Australian Computer Music Conference 2009*, ed. Andrew Sorensen (Melbourne: Australasian Computer Music Association, 2009), 1–16.

[47] Galloway, *The Interface Effect*, 86.

very useful as a way into naming the fundamental yet hidden phenomenon that confronts us at digital interfaces:

> Algorithmic interfaces – even as they flaunt their own highly precise, virtuosic levels of detail – prove that something is happening behind and beyond the visible. In other words, *there are some things that are unrepresentable*. And the computer is our guide into that realm.[48]

While we understand that there is much then that we as ordinary users cannot apprehend about the raw data shaped and represented in Google Earth, we can, however, use our awareness that Google Earth consists of a mass of algorithms as a basis for understanding the platform, and turn our attention towards the textual content of the interface itself and how it addresses us – how it *performs* us.

So what do we see on our computer screens when we open Google Earth?

Figure 2.2 is a screen capture of what I saw when I opened Google Earth on 1 November 2013, using Version 6.1.0.5001 which was enabled

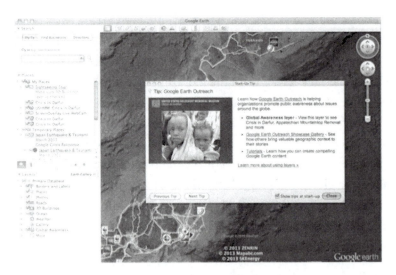

Figure 2.2 Screenshot from Google Earth © 'my first page to open' 1 November 2013, 3.38 p.m. EST. Google and the Google logo are registered trademarks of Google Inc., used with permission.

---

[48] Galloway, *The Interface Effect*, 86.

on 17 October 2011. There is clearly a lot of information presented first up, and quite a proportion of this is on ways to use and access the 3D modelling and animations that are available as images in Google Earth.

Icons at the top of the page enable links to various viewing options, including links to

- Google Maps: street maps, street views and 3D imaging of buildings via aerial photography and 3D modelling;
- views of 'earth, sky and other planets';
- the passage of the sun across landscapes as seen at particular times;
- views of landscapes that can show the passage of time, via images of a place at different times of image acquisition.

There are also icons that help you email the page, print the page, tag it, place markers on the page and create layers. Other icons can link you to processes such as creating a journey between places and recording that journey. There is an important icon at the top of the page that when clicked on shows a toolbar. In this toolbar there are many more viewing options, including the various layers that are available – functions that will plot a journey between places, and a section called 'My Places'. In this latter section you can archive places and layers you are interested in and want to return to, as well as recordings you might have made of specific journeys. In Figure 2.2, my 'My Places' includes the Global Awareness Layers of sites that I have researched: for example, Crisis in Darfur and the March 2011 Japanese Earthquake and Tsunami site. Each time you open Google Earth, you are also presented with a series of windows called 'Start-up Tips'. These cycle through each opening and by chance, the one that appears in Figure 2.2 is about the Global Awareness Layer called Crisis in Darfur: the layer that constitutes my case study for this book and which I analyse in Chapters 4 and 5. The photograph 'Two Sisters' that appears in this window is actually also one of the particular images that I focus on in Chapter 5. However, the aesthetics of the whole page as it opens is announced (almost as an opening of curtains on a proscenium stage) by a blue globe spinning out of black space, coming closer and bringing into view whatever map

Google Earth has remembered from your most recent image repertoire in My Places. Immediately, the Earth itself is customized for your own personal use.

Layers

Before moving on to the more detailed discussions, in both this and later chapters, of Google Earth layers and how they can be customized, it is useful to note how the term 'layer' is used as a way to describe an interactive mapping function in Google Earth. The term has a broad application in digital visualization technologies. In an introduction to his series of online tutorials on the graphic design raster editing programme Adobe Photoshop, Colin Smith describes a layer as 'simply one image stacked on top of another'.[49] Introduced in Photoshop 3.0, the use of layer panels allowed designers who were not coders to manipulate images according to the properties chosen for the layers they created. Smith lists several kinds of layers, including Thumbnails, Layer Groups, Type Layers, Adjustment Layers, Layer Masks and Smart Objects 'that can hold; multiple (or 1) layers, vectors for illustrator, raw files, video, 3D or many other types of objects'.[50] Whilst these definitions are made in the context of Photoshop, they are also basically applicable to layers in Google Earth in the sense that they can be used to make available added information to be included in the graphic design production of Google Earth.

In general terms, Google Earth Layers are simply, in David A. Crowder's words:

added pieces of information above and beyond just the satellite image itself. In fact, everything in Google Earth, except for placemarks, that isn't a photo from space is a layer of some kind.[51]

---

[49] Colin Smith, 'Photoshop Layers, 101', http://www.photoshopcafe.com/tutorials/layers/intro.htm (accessed 8 August 2014).
[50] Colin Smith, 'Photoshop Layers, 101'.
[51] David A. Crowder, *Google Earth for Dummies* (Indianapolis, IN: Wiley Publishing, Inc., 2007), 65.

The layer that we perhaps identify most strongly with Google Earth is the Terrain Layer with 3D information obtained from NASA's Shuttle Radar Topography Mission,[52] and in the Google Earth '7+ free client' platform this layer cannot be turned off. But every layer's accuracy and currency of information, whether created by Google Earth or its users, always relies on the data supplied. Google Earth relies on the trustworthiness of the various data sources it licenses from or purchases. In Chapter 3, I discuss the social and scholarly implications of digital mapping projects, but here it is useful to note that Google offers free online tutorials showing how both vector and raster maps can be imported into the Google Earth platform. There are various levels of competence offered, varying from individual users creating simple layers to show the path they walked around a mountain, to complex models of how cities might grow – for marketing and real estate development projects. Information and tutorials are provided at the level of software developers and at the level of everyday non-coding users.

In locations mapped by Google for Google Maps and Street View, these maps and views can now be incorporated within viewing windows of Google Earth; and Google Earth images can now be included and layered over Google Maps. The entry 'Viewing Data from Google Earth' in the Google Support forum describes and shows how this latter process is possible:

> Google Maps can now read KML or KMZ, the file formats Google Earth uses for the exchange of geographic information. This means that you can view data you create and share with Google Earth on Google Maps. You can use this feature to plot multiple points of interest, draw lines, and mark regions on Maps.[53]

In other words, raster, digitally sourced image layers of information created in Google Earth can now be layered onto the vector-created maps of

---

[52] See Wikipedia 'Google Earth', http://en.wikipedia.org/wiki/Google_Earth (accessed 8 August 2014).
[53] 'Viewing Data from Google Earth', https://support.google.com/maps/answer/41136?hl=en (accessed 8 August 2014).

Google Maps. Similarly, vector-created maps can be imported into a view of Google Earth and can thereby include various 'smart objects' containing further embedded information layers. The Google Earth toolbar now also allows a simple flip into a Google Maps view of the same location.

The 'My Places' area on your home page in Google Earth will show layers you have created yourself and other layers that you have visited and saved. There are also a plethora of layers that Google Earth itself makes available, including national border layers, advertising layers, weather, cloud cover, seabed viewing layers and so on. The layers can be turned on or turned off with a click of the mouse, but the challenge sometimes is how and when to turn off a layer. When many available layers are turned on, either inadvertently or intentionally, the viewer can be confronted with a mess of imagery and information that masks and distracts from the information originally sought. Figure 2.3 demonstrates a confusion of information (even without the weather layer) that resulted from my search for information on tsunamis and earthquakes in Japan.

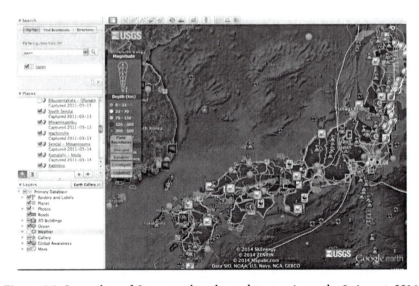

Figure 2.3 Screenshot of Japan earthquake and tsunami search, 8 August 2014, 6.09 p.m. EST © Google Earth. Google and the Google logo are registered trademarks of Google Inc., used with permission.

In a conceptual and perhaps historical sense, the digital technology of layering images can be sourced back to the traditional analogue process of cel animation,[54] together with all the other magical devices of illusion that humans have used for manipulating sight, for the many and various purposes of pleasure, knowledge and deception.

Embodied space at the interface

Clearly Google Earth was from the beginning and still is a textual combination of the imagined, the virtual, the material, movements between actual times and spaces and all the actual and potential interactions in between. As my research moved into the stage of finding out about Google Earth's infrastructure, I found myself wanting to be able to directly reference the orbiting satellites which capture the digital data that is coded then into the wondrous images and interactive processes that constitute the site. My main reason for wanting to do this was that my imagined Google Earth, my awareness of Google Earth, included the presence of these remote sensing satellites that orbit Earth, the data collection centres on Earth and the elaborate coding that is used to produce the site's composite images. In other words, my imagined 'virtual globe' included not only its users and producers (who are sometimes the same people) but also what I thought of as an elegant array of sophisticated machinic bodies in space. As a non-coder who admires the skills of those who do code and the skills of space technicians and space explorers, I was initially and easily drawn into the romance of satellites as machines

---

[54] See Annette Kuhn and Guy Westwell (eds), *A Dictionary of Film Studies* (Oxford: Oxford University Press, 2012), online version 2014, eISBN 9780191744426: 'A traditional animation technique whereby a transparent sheet, or cel (celluloid), is used for each frame. The cel contains the moving elements of the image, with the animator making small incremental changes from one cel to the next. As the animation is photographed, the transparent cel is placed over a drawing containing any non-moving elements that recur from frame to frame, thus preventing the need to repeatedly redraw these elements. As the image is photographed, the background and cel are synthesized, and when projected this creates a singular image and the illusion of movement'. http://www.oxfordreference.com.virtual.anu.edu.au/view/10.1093/acref/9780199587261.001.0001/acref-9780199587261-e-0095?rskey=jSjHHn&result=94 (accessed 10 August 2014).

that can 'see' into space: machines that can envisage for us our spinning planet as a whole, in all its 3D glory – as made available through aerial photography and remote sensing satellite data.

As a way of grounding my speculations then about Earth as shown via Google Earth, I wanted to understand a little better the satellite technologies that play such an important role in its production. As part of my plan to close to some extent the gap noted by Parks and Schwoch, that knowledge gap between me and the people who produce the platform, I even posted a naïve question to the Google Earth Product Forum within the Google Earth Community, asking 'what satellites and satellite owning corporations is Google Earth using currently?' I got one reply: 'ALL OF THEM, mwahahaha!' As good an answer as any perhaps, although probably an overstatement: there are currently over 500 communications satellites orbiting Earth, and more than 300 of them commercially operated, 10 per cent are remote sensing satellites.[55]

Much information about satellite data collection actually sits at the bottom of Google Earth images, and although a survey of this information would be an interesting addition to any historical snapshot of Google Earth technology, it is the extent of access that Google has to satellite data that is perhaps most significant. Certainly, the huge image database of the private corporation DigitalGlobe[56] is the primary database for Google Earth images, and as their web page currently displays, they use information from five high-resolution satellites. The current resolution for the GeoEye-owned IKONOS-2 via synthetic aperture radar is 1 m spatial resolution.[57] This information on current satellite technology is quite accessible via a web search, together with who actually owns the satellites and how you can purchase specific time/space visual imagery via DigitalGlobe. The processes of how

---

[55] The Tauri Group for the Satellite Industry Association (SIA), 'State of the Satellite Industry Report, June 2013'.

[56] See DigitalGlobe's own 'overview' of its business concerns: 'DigitalGlobe owns and operates the most agile and sophisticated constellation of high-resolution commercial Earth imaging satellites. IKONOS, QuickBird, WorldView-1, Google EarthoEye-1 and WorldView-2 together are capable of collecting over 1 billion km² of quality imagery per year and offering intraday revisits around the globe. Add to that our aerial program offering wall-to-wall coverage of the U.S. and Western Europe'. http://www.digitalglobe.com/about-us/content-collection-overview (accessed 31 October 2013).

[57] ESA Earth Net Online, https://Earth.esa.int/web/guest/missions/3rd-party-missions/current-missions/ikonos-2 (accessed 5 November 2013).

this imagery is sensed, collected and digitally processed are also available in undergraduate and graduate textbooks and other scholarly writings.[58] In April 2013, DigitalGlobe bought the GIS intelligence crowdsourcing company Tomnod. As described in the DigitalGlobe blog, 'Tomnod has been at the forefront of innovation in the growing field of crowd sourcing of earth observation imagery analysis, combining their unique and novel algorithms with deep GIS and imagery knowledge.'[59] DigitalGlobe thereby extended the accessibility of its remotely sensed real-time imagery directly and at no extra cost to the general public.

The difference between Tomnod and Google Earth is that Tomnod can offer maps with more real-time capability than Google Earth. This difference is illustrated very clearly by a tragic event that was unfolding at the time of writing. On 8 March 2014, Malaysian Airlines flight MH370 from Kuala Lumpur in Malaysia to Beijing, China, disappeared from the screens of air traffic controllers less than an hour after take-off. The plane's transponder had been somehow turned off but primary radar that was accessed later showed that the plane had continued to fly for 7.5 hours after its disappearance. The loss of this international airliner, a Boeing 777-200ER carrying 239 people, drew the worried attention of governments and people all round the world. Many nations are currently involved in a search that came to focus on a huge flight corridor in the Southern Indian Ocean, based on closely studied radar information. Some satellite images showed up images that suggested large pieces of plane wreckage in an area over 2,000 km southwest of Perth, Australia. Tomnod set up the customized search site www.tomnod.com/nod/challenge/mh370_indian_ocean/map/, while Google Earth warned people not to use Google Earth as it was not a site for images in real time.[60] Entry into the Tomnod search site for MH370 is via a

[58] See, for example, Brinda M. Chotaliya, Sarang Masani, 'Remote Sensing: Essentials and Applications', *International Journal of Engineering Trends and Technology* 4, no. 8 (2013): 3460–3467. ISSN 2231–5381 and John A. Richards, *Remote Sensing Digital Image Analysis. An Introduction*, Fifth Edition (Berlin and Heidelberg: Springer Publications, 2013), doi: 10.1007/978-3-642-30062-2.

[59] DigitalGlobe, http://www.digitalglobeblog.com/2013/04/08/tomnod/ (published 8 April 2013, accessed 22 March 2014).

[60] K. J. Akash, 'Missing Malaysian Airline Flight MH370: Don't Use Google Maps to Search for Plane, Says Google', *International Business Times* (11 March 2014), http://www.ibtimes.co.uk/missing-malaysia-airlines-flight-mh370-dont-use-google-maps-search-plane-says-google-1439744 (accessed 22 March 2014).

link (accessed 22 March 2014) on the home page: http://www.tomnod.com/
nod/. When you join the search by clicking on the 'join now' link, you go to
a site at http://www.tomnod.com/nod/challenge/mh370_indian_ocean. The
search website does look similar to a gaming website, with three buttons
showing you variously how many map tiles you have explored, how many
items you have tagged and how many people agree with your tagging. This
search site is very much a community site with the combined aesthetic of
serious GIS mapping analysis and gaming; as suggested by the URL, the site
offers 'a challenge'.

What becomes important about the combination of web design and
GIS displays that constitute Tomnod and Google Earth is that it eventually
produces an interface with an aesthetic that disseminates and interprets factual
GIS information. In the case of Google Earth, this interface shows models and
animations that are used by so many people for so many different reasons,
in so many ways and to obtain so many different kinds of information. As
the search for MH370 illustrates, the crowdsourcing opportunities offered by
Google Earth, and although to a lesser extent Tomnod, can themselves invite
a simplistic, reductive attitude in users as to what can be seen via amateur
analyses of satellite imagery.

I argue, however, that both broadly social and individual imaginaries
(even to some extent, the more romantic imaginary we might call 'general
optimism') form a valid as well as significant component of sites such as
Tomnod and Google Earth. In this context then, I disagree with Baudrillard's
quite pessimistic conceptualization of the simulated hyperreal: what the
real has turned into as a result of algorithmic modelling. He claims that 'the
cartographer's mad project of the ideal coextensivity of map and territory'
disappears in the simulation and the miniaturization that constitutes speculative
models of imagined actuality. Simulation in his sense is a model of something
that cannot exist even as an ideation (of the mad cartographer and many
others).[61] He doesn't call this kind of simulation 'not real', but 'hyperreal'. Such
an idea of the hyperreal begs, however, the question of modelling processes
and objects that invite speculation/imagination as modes of interpretation. I

---

[61] Baudrillard, *Simulacra and Simulation*, 2–3.

propose that Baudrillard's concept of the hyperreal loses some of its force at this current moment in time. Now we are once again looking into processes of representation for glimpses, for the fleeting, changing signs of what actually exists 'beyond representation'. The very use of the term 'actual' in scholarly and journalistic texts is a telling indication of how we are currently looking for ways to sidestep issues and studies of 'the real' that have gone before.

Currently, we talk about the virtual and the actual, and say both of them are real! A questioning of what is real or not has come a long way in cartography since the advent of digital mapping. In light of philosophers' continuing debates on the relationship between the representational processes of simulation and the actual world, it is clear that the general public that consists of amateur cartographers must still be grappling with (and/or ignoring) distinctions between algorithmic modelling and the space that is being modelled. The development of crowdsourcing in cartography informs us about a new social imaginary that draws both on the delight in imaging Earth and a sense of democratization, a weakening of the power relations that favour people who know how to hack and people who don't. At the same time, crowdsourcing also serves to emphasize this knowledge divide when ignorance becomes a constraint on acts of making sense of information.

This component of the imaginary then needs to be balanced against the prosaic details of those power relations involved in the pragmatic production of Google Earth as well as the relations embedded in the potential for the exploitation of less IT literate and so perhaps/probably less empowered users of the site. The empowerment I am talking about here is not simply one derived from knowing how the algorithms are being used to shape information. There is also the power quotient that can come from expanded knowledge: of having the ability to gain even more information from the site because you know how it got there (even if you don't know the details of this process) and therefore you know more about how to interpret the site and its images for your own purposes. There is, however, another set of people who can be disempowered by Google Earth, and again, these people are also us.

We live on planet Earth. When our planet is represented, to some extent, so are we represented. We are being looked at. These images, visual information derived from data supplied by Earth-observing media (raw data shaped into information), also invite then this question: what and who actually exists on this planet? Who and what are the living sensory bodies and geographies that are inscribed by Google Earth image making and using experiences? Who gains? Who loses?

To say that 'we are being looked at' is also to say that we are subjects of surveillance by other humans or by machines that other humans can use to look at us. Before moving on (as I do in Chapter 3) to consider how mapping and surveillance coincide to produce one of the key visible aspects of Google Earth, I need to reintroduce the human body into this discussion – in the context of how it has been thought of as a site for sensory perception of the world and also as a crucial actor/factor for defining space. In *The Phenomenology of Perception*, Merleau-Ponty works together ideas about the actual existence of our bodies as perceptive entities that are also defined by their inscription and of the world through co-relative acts of perception: 'My body is that meaningful core which behaves like a general function, and which nevertheless exists, and is susceptible to diseases.'[62] In this short sentence, he describes the ways the human body is inscribed with cultural meaning, how it has an actual existence as something that can work well or break according to its physical environment, how it can have an effect on the world in ways that are largely dependent on acts of perception by itself and others. For the purpose of interrogating a communicative cyberspace object such as Google Earth, it is the notion of space in relation to the human body that becomes very important for understanding questions of power and human agency. Using Merleau-Ponty's words again: '... the experience of our own body teaches us to embed space in existence.'[63] How does this embedding happen?

Drawing on Merleau-Ponty's theory of perception and Henri Lefebvre's work *The Production of Space*,[64] Jason Farman has developed his own definition of embodiment and space for using in his description of the mobile digital interface. He quotes Lefebvre's description of space as a social reality,

---

[62] Maurice Merleau-Ponty, *Phenomenology of Perception*, trans. Colin Smith (London: Routledge, [1945] 2004), 170.

[63] Merleau-Ponty, *Phenomenology of Perception*, 171.

[64] Henri Lefebvre, *The Production of Space* (Oxford: Blackwell Publishing, 1991), 170.

a spatial concept that is never empty, that cannot be thought of as a container of anything because it can only be defined via physical and perceptual relationships.[65] Farman's conceptualization of the body as embodied through space is an intricate journey through complex theories. His concluding definition of what he calls a 'mode of phenomenological hermeneutics' is extremely useful for describing how we can think about our own subjective status in the Google Earth experience:

> We are embodied through our perceptive being-in the-world and simultaneously through our reading of the world and our place as an inscribed body in the world.[66]

I continue this discussion of embodied space in Google Earth in Chapters 4 and 5. In these chapters, I look closely at both effects and affects that can be identified as the result of specific relationships occurring in online human rights activism. For now, I want to continue a more general discussion of how we can think of Google Earth as a digital object with an aesthetic that uses animated images of Earth and brings them to us 'for free'. Google Earth's basic service 'for free' is in the context of a satellite industry that showed total global revenue in 2012 of USD189.5 billion and global revenue for remote sensing satellites at USD1.3 billion.[67] And in the case of Google.com as a whole, this 'for free' service comes in the context of advertising, for-fee investment and development services and the carbon and other ecological footprints implicated in the operations of a huge international corporation (no matter how such a corporation describes itself or aspires/tries to be).

The Google Earth platform, to a large extent, has always been based on an embodied aesthetic of naïve vision and touch; it implicates our eyes, the touch of our fingers on a screen or a mouse and keyboard and it offers the haptic experience of imagined flight. Google Earth provides an instance of Haraway's 'god-trick', the illusion of infinite vision, available to download for

---

[65] Jason Farman, *Mobile Interface Theory. Embodied Space and Locative Media* (New York, NY: Routledge, 2012), 18.

[66] Farman, *Mobile Interface Theory*, 33.

[67] The Tauri Group for the Satellite Industry Association (SIA), 'State of the Satellite Industry Report, June 2013'.

no financial extra cost via most web browsers (although for a while Google's web browser Chrome was by far the easiest to use for downloading Google Earth). It even contains 'time-line functionality allowing elements with temporal information to be encoded (KMLv2.1) and then selected and filtered by the user'[68]: 'data can also be streamed from a server in response to changes in the visible area of the viewing window sent by Google Earth.'[69]

Not surprisingly then, my initial approach to the aesthetic of Google Earth can perhaps best be understood as 'Dionysian' in the sense ascribed by Paul Kingsbury and John Paul Jones III in their discussion of Google Earth as 'the projection of an uncertain orb spangled with vertiginous paranoia, frenzied navigation, jubilatory dissolution, and intoxicating giddiness'.[70] But Munster's equally valid description of Google Earth's virtual globe takes this celebratory vision down a darker path of alienation into what she calls 'Google Earth's reterritorializing cartographic quest'[71]:

> Google Earth is just that distant planet that seems strangely suspended from the chaos of sociality and life.[72]

For all its splendour together with the possibilities for intimate exploration that it offers, Google Earth nevertheless also visualizes Earth as seen from above, as isolated in space, as distant, as Other.

As noted earlier, Google's history of its own development states that its launch of Google Mapmaker in 2008 helped ensure that 'the map accurately reflects the world'.[73] The map in question comprises the collection of maps that can emerge from within Google Maps and Google Earth. In view of the spectacular attractions[74] and sensory affects of its visualizations and

---

[68] Wood et al., 'Interactive Visual Exploration of a Large Spatio-Temporal Dataset', 1177.
[69] Wood et al., 'Interactive Visual Exploration of a Large Spatio-Temporal Dataset', 1177.
[70] Paul Kingsbury and John Paul Jones III, 'Walter Benjamin's Dionysian Adventures on Google Earth', *Geoforum* 40, no. 4 (2009): 502–513.
[71] Munster, *An Aesthesia of Networks*, 46.
[72] Munster, *An Aesthesia of Networks*, 46.
[73] http://www.google.com.au/about/company/history/ (accessed 2 November 2013).
[74] This important theory in film reception theory is drawn from Tom Gunning's seminal writing on early cinema: 'The Cinema of Attractions: Early Film, Its Spectator and the Avant-Garde', *Wide Angle* 8, Nos. 3 and 4 (1986): 63–70. The primary idea is that people will go to see a film that offers high levels of audiovisual pleasure (including that of horror) in an increased or new way. So novelty and sensory affect play a large factor in why a not-so-profound film appeals to an audience. Early film snippets were sometimes viewed in booths on amusement piers.

animated modelling, Google Earth extends this cartographic mission towards an enterprise that offers an affect of 'being the world'. How does this affect of being in the world tally with Google's mission of accurately reflecting the world via Google Earth? And how does it tally with the sense of isolation described by Munster?

In Baudrillard's terms as noted earlier, there is no tally at all: there is only a modelling process that disguises and so cancels out recognition of the actual. Yet there is an undeniable affect to be experienced when using Google Earth. It is not simply the affect born of engaging with a fantastical world (such as Baudrillard's example of Disneyland in California, United States). There is an inherent sense of mystery or alchemy embedded in the realm of the imaginary that is required in order to experience Google Earth. I am not using these terms to obfuscate the ethical and philosophical issues raised by the site, but rather to suggest that an isolated critique of power relations implicated in Google Earth can run the risk of obscuring the degree of narrative agency that can be achieved through how we address the phenomenon that is Google Earth.

To some extent, such a sense of mystery can also be felt when using any technology that the user does not understand; the user well knows that there are 'experts' (the coding 'priests' of IT) who have created the technology and who usually know how it works and how to fix it when it goes wrong. I suggest though that a site such as Google Earth, which channels information gathered from satellites orbiting the Earth to the everyday non-coder user, has a much greater degree of charisma than other sites, and hence a greater capacity for seeming mysterious. This charisma comes from the way mysterious (to the non-coder) algorithms nevertheless seem to give your ordinary person the vision of an astronaut orbiting in space above the Earth. We can literally use, together with NASA, images from space to tell our own stories, even if there is always the possibility that at the same time we are being manipulated, conned by Google into thinking that this corporation and its many products are the natural or even the only digital vehicles for accessing the social and physical world. In Google Earth, we already have cultural stories of astronauts and space travel. These stories to a degree mitigate Munster's image of an isolated planet disengaged with its human inhabitants.

There is always the likelihood that we currently are prone to the 'googlization'[75] of how we understand the world around us, and Munster contrasts 'the conjunctive relation between distributed collectivity and cloying intimacy [which] as vital to networked experience'[76] with what she maintains is Google's 'solipsistic quest to make the world "googley"'.[77] She goes on to say that this quest is 'unachievable', and closely examines the ontological state of 'autopoiea' that 'googlization' infers.[78] I certainly agree that such a quest is unachievable. The quest itself can be thought to describe the very existence of an impossible vanishing point defined by the gap between how Google might imagine its own possibilities and the constraints of actuality: what Munster calls a 'holding apart of Google and the world [that] opens up just enough of a crack in the ground to ward off self-enclosure'.[79]

I propose that such an inherent (and to be hoped for) failure is due more to the potential for a process of subversion that is necessarily offered in the context of Google Earth than in the context of Google's basic function as a search engine, which offers a finite number of search parameters. Google Earth 'screams out', so to speak, for personalized, individualized usage, of feeling and knowing *'life as search experience'*.[80] It is not a Facebook-like platform that provides a very constrained website design for presenting an online identity whilst simultaneously enabling maximum exposure of that identity. It is not a YouTube-like platform whose main purpose is for downloading and uploading personal and/or public videos of place, self and other people. Google Earth offers personalized virtual journeys through space and time and ways of tracing and archiving those journeys. It invites

[75] Munster, *An Aesthesia of Networks*, 46. See Siva Vaidhyananthan, *The Googlization of Everything (And Why We Should Worry)* (Berkeley, CA: University of California Press, 2011).

[76] Munster, *An Aesthesia of Networks*, 47. Munster attributes this 'conjunctive relation' to Natalie Bookchin's approach. See Natalie Bookchin and Blake Stimson, 'Out in Public: Natalie Bookchin in Conversation with Blake Stimson', in *Video Vortex Reader II: Moving Images Beyond YouTube*, eds Geert Lovink and Rachel Sommers Miles (Amsterdam, Institute of Network Cultures, 2011), 306–317.

[77] Munster, *An Aesthesia of Networks*, 47.

[78] Munster, *An Aesthesia of Networks*, 45–71. Munster examines Google Earth within the context of an emerging 'networked ecology of "Google-us"' (46), a context that focuses on Google as a search engine.

[79] Munster, *An Aesthesia of Networks*, 47.

[80] Munster, *An Aesthesia of Networks*, 61.

us, via its toolbox of options for using the site, to tell our own stories. I suggest that this invitation is to a much less controlled set of options than the timelines, self-promotion and 'friending' available via a site like Facebook.

Rather than working within the constraints of seemingly tightly controlled parameters of chosen identity, the Google Earth mapping platform simply provides some tagging tools and viewing options for us to try and create a personal view of Earth – we are given access and some degree of control over how Earth is represented both to ourselves and to others. This control is to a degree delusional of course, because it is always reliant on the algorithms and the people who create software from them. Nevertheless, we can create stories that pit the one (of us) against the enormity of the world. These stories might not be easily visible to others (unless we make them so via accessible layers) but they can be very important to the ways in which we understand ourselves in relation to our environment.

So much information

In today's vernacular, to 'google' is to search the WWW. Usually, to say that we are searching is to say that we are searching to find answers; but web surfing can simply be 'riding the wave' of carefully ranked data, searching for the joy of it – the haptic synesthesia of mind gliding through ideas and putative answers to questions hardly formulated. In Google Earth, we search our planet to find out things that we did not know and to tag what we think we know about a location, to a map of that location. We can put the mark of our own presence onto the mysterious unknown of the Earth – that planet which we and our satellites cannot see or comprehend as a whole in real time. Google Earth is an interesting site because its aesthetic so powerfully creates and shapes questions we might not have known we wanted to ask.

The primary way to use Google Earth is for individualized exercises in surveillance of places, events and landscapes. The use of the Google Earth Community also explicitly predicates an experience of that 'cloying intimacy' noted by Munster as existing in network ecologies. Cloying or not, however, the primary intimacy that plays a role in the phenomenon of Google Earth is not one that is necessarily exposed to a network of online mass audiences. The intimacy is born of a particular kind of individualized searching. Its parameters are continually expanding with every update and upgrade, as noted on links to the website itself. We can fairly easily customize the parameters of our searching, and in ways that the site makes transparent (though degrees of transparency do depend on IT literacy). Google Earth then offers as part of its experience a possibility to subvert to some extent the knowledge hegemony created by the existence of ever-increasing and seemingly infinite digitized databases. It rather offers a new hegemony of both actual and imagined individual agencies. This new hegemony is embedded within other social media to some extent but in the case of Google Earth, it manifests more reliably as an attainable goal. So while offering the data that is created via the older hegemony of 'science' – only the trained, privileged few can access knowledge derived from satellite technology – Google Earth also relies on/needs users to look for ways to personalize knowledge in order for it to complete its active presence as a particular kind of social or socialized, personalized media.

After opening Google Earth, my own reactions to the first page usually include anticipation, frustration and some confusion. These reactions are the result of an obvious but challenging aspect of Google Earth (both for Google and for us the users): the staggering amount of information and ways of viewing this information that it offers the user. As Google says about the whole Google.com enterprise:

> The name – a play on the word 'googol', a mathematical term for the number represented by the numeral 1 followed by 100 zeros – reflects Larry Page and Sergey Brin's mission to organize a seemingly infinite amount of information on the web.[81]

The challenge for Google is to offer simultaneously this huge amount of information and to present it as a nevertheless accessible and easily

---

[81] http://www.google.com.au/about/company/history/ (accessed 2 November 2013).

imagined package of information. Google presents Google Earth as a truthful viewing platform for planet Earth. It is now one of several 'Earth observation' software platforms. How do we as ordinary everyday users respond then to this enormous package amount of visual information that it can be so easily accessed and used by anyone familiar with using Web 2.0?

I suggest that a significant affect derived from the Google Earth experience is one born of a particular kind of confusion that comes from combining the *jouissance* of an immersive cyber experience with the demanding task of extracting data from a complex website. In other words, this affect is offered/potentiated by the combination of a particular experience and a specific task. If such an affect can be described in this way, then it is also possible to consider it as another way in which Google Earth paradoxically manifests to us as a vehicle for research, communication and interpretation, all at the same time. It induces us to claim: 'We have the power (to see to know)!' when all along we the users and makers of the site are simply playing with possibilities of knowledge via the mathematics of algorithms and the biology of vision. Google Earth explicitly and simultaneously invites both speculation and imagination and provokes that part of the human that wants to 'be sure about things'. Always a dangerous combination! At the same time, Google Earth plays an important, unique and celebratory role, in McClendon's words, 'not as a digital record of what [Earth] was, but instead an evolving symbol of what will be'.[82]

---

[82] McClendon, 'UNEP Award speech'.

3

Using Google Earth – Who, How and Why

For all its armored solipsism, the digital pretends that because it is
unified it is unique, and because it is enclosed in its own sphere, there
are no far horizons. There are, and they are plural.

Sean Cubitt[1]

All the people who use this digital globe and its affiliated services create Google Earth's horizons. The interactive design of Google Earth allows viewers to feel control over what they are seeing, to tag places significant to them and to join in a community of other enthusiasts. A viewer can save her 'own' version of Google Earth, as if it were another Web 2.0 personalized social networking website. Our perceived conflation of time and space when viewing the images found in Google Earth results in a simulated form of immediacy in communication. Our sense of the 'far away' is particularly affected. As individuals, we view Google Earth on our own small, personalized computer screens: the same screens through which we immerse ourselves in private correspondence. I suggest that we interact with Google Earth with a similar sense of immersion and that our engagements with people and places through Google Earth are embodied and emotively charged. They exist not only in the simulated world of 'as if' existence in the 'virtual space' of software applications, but alongside our other active, bodily engagements with the world around us. Our immersion into this 'as if' world contributes to the 'armored solipsism' of digital technology as it manifests in Google Earth. If we undertake to use

---

[1] Sean Cubitt, *The Cinema Affect*, 272.

Google Earth we are also undertaking a whole raft of assumptions about Google Earth itself. These assumptions include: that it is useful to our purposes, that our immersive, almost intuitive kind of interaction with its screen content is necessary in engaging with it and that when we use Google Earth, we are engaged in a serious discourse – about how the world is. This chapter interrogates the motivations and manipulations inherent to these assumptions by asking three questions: Who uses Google Earth? Why? And how? It is not difficult in the first instance to answer these questions. Google and Google Earth web pages provide us with a lot of information that we can use for this venture. Research into the Google Earth Outreach programme, founded by Rebecca Moore, shows how much effort Google Earth puts into engaging with communities of scholars, government bodies, businesses, not-for-profit (NFP) organizations and individuals and people with special interests. In this chapter, I look into the Google Earth Outreach programme to illustrate the wide range of people who use the Google Earth, their purposes and how Google Earth enables their use of Google Earth for these purposes.

First, it is useful to revisit the most easily understood possible outcomes of using Google Earth: we can find maps, search maps, annotate maps and even create customized maps. Put simply, mapping is the most obvious outcome of using Google Earth. Google Earth's visual modelling of planet Earth via information from remote sensing of satellites in orbit around it clearly draws on a particular cultural performance that is inherent to humanity itself: cartography. This performance has changed dramatically in the digital era, as noted in Chapter 2. As with Google Earth, most current cartography relies on satellites orbiting Earth. In Parks' and Schwoch's words:

> As machines that orbit our planet, satellites are uniquely positioned to visually represent Earth, and their images have been composited to construct earth as 'whole'.[2]

---

[2] Parks and Schwoch, *Down to Earth*, 3.

What is interesting, however, is that Google Earth's adoption of visually articulated mechanisms for outreach and activism actually contradict or at least break down the sense of wholeness, unity or sameness that is created by representations of the blue globe depicted so often and in so many contexts. Tagging of spaces shown on Google Earth personalizes, even to some extent individualizes, times and spaces that constitute our personal awareness of our environment as part of planet Earth. Whether this tagging is via someone's terrible photo of some tourist destination or the more sublime tagging that is the Global Awareness Layer of 'Crisis in Darfur' on the map of Africa, the globe is literally tagged as being part of people's memories, experiences and endeavours. The embedding of Google Maps onto Google Earth also introduces the ubiquity of Google mapping enterprises and how views from Google Maps are now commonly linked and actually embedded in many other websites.

The kinds of mapping offered by Google Earth provide a form of spatial analytics that can be used by institutions to easily communicate, even to the general public, information about specific locations, as they currently exist in time and space. This chapter describes some of the ways in which governments, businesses and research organizations use Google Earth. An account of how activist organizations use Google Earth and/ or have been influenced by the Google Earth platform to use satellite technologies that are provided by other corporations is included in Chapters 4 and 5.

Many different groups of people use the Google Earth site. Reasons for this use range from agricultural and environmental research to corporate advertising to human rights activism to environmental activism to searches for traces of extraterrestrial presence on earth to military and corporate surveillance activities. And then, as noted earlier, there is perhaps the most prevalent use: individual people's mundane uses of Google Earth to plot and tag their own journeys through landscape, to recognize familiar terrains from a 'god-like' perspective and to access information that is directly linked or tagged to a particular place on Earth.

My discussion of 'who' uses Google Earth and 'why' necessarily draws on the 'how': how Google Earth promotes, provokes and invites, purposefully or not, our participation in this particular instance of digital mapping. Purveyors of geographical information systems (GIS) such as Google Earth have introduced into the public realm a form of mapping space/time that is simultaneously localized and global. You can focus close in to a particular place and time, and then you can draw out to view the existence of that place/time as part of Earth's geography as a whole – and you can do this by simply sliding your finger over the top of your computer mouse. In Chapter 2, I presented some of the ways in which Google Earth offered us as individuals personalized mapping and virtual journeys between locations imaged within the platform itself. Before looking at the implications of this kind of personal mapping and journeying, I want to consider here one kind of usage that comes under the rubric of Google Earth Outreach.

This kind of usage is usually undertaken by a group of people who interact with each other online with some kind of common purpose. This new kind of social grouping results in what web analyst and social researcher Robert Ackland defines as an *online group*: 'a group of people who conduct personal computer-mediated interactions, where interaction is focused on a topic that reflects the common interests of the group'.[3] However, he also says that a collective identity such as exists in what is described as a virtual or online community does not exist in an *online group*.[4] This is an interesting distinction when we consider Google's nurturing of the Google Earth Community site – through which Google appears to want to create some kind of collective identity between people who have no other interest in each other than a shared use and fascination with the Google Earth platform. The online groups that I describe in this chapter comprise people engaged not just in a topic of common interest but in a shared task. In other words, an online group therefore might be a specific group of people who are tasked with collating and disseminating

---

[3]   Robert Ackland, *Web Social Science. Concepts, Data and Tools for Social Scientists in the Digital Age* (Los Angeles, CA: Sage, 2013), 11.

[4]   Ackland, *Web Social Science*, 11.

data for an organization whose task is not necessarily one that exists of itself online. In the context of groups who use Google Earth, this data needs to be in a visual format in order for it to be useful.

'Dual use' of data visualization

The visualization of digital data has been an important process for understanding what the data means. Again, this idea that visualization of information is important for understanding any kind of data is not in any way new. What is new is the technology and the ways data can be visualized. Visual artists have long been interested in the aesthetics of visualizing digital data,[5] and so have the military. For instance, in the Preface of a report prepared for NATO by its Research and Technology Organisation in 2001, we find an interesting definition of visualization in the context not only of analysis but also of human perception.[6] The authors of this Report use the verb 'sense' rather than 'see' in order to describe how visualization occurs as an inclusive, pan-sensual kind of perception of the world.[7]

The idea of visualization as an act that draws on more than the sense of sight is a highly significant one when thinking about Google Earth. This is because when we use this particular GIS, we do bring to our experience of searching through images of Earth, all the hermeneutic prejudices that we bring to any kind of text that we might try to understand. We bring our memories, real and imagined, our speculations on the future, our overall health and our overall literacy in the technology used to create the text. All of these entities, including the text itself, can be primarily located within the domain of senses other than sight. They include touch, hearing, smell and time and spatial awareness. The fact that military organizations acknowledge the subjective position of an analyst, as do analysts in the academic

---

5   See, for example, Munster, *An Aesthesia of Networks*, for a broader look at recent conjunctions of art and technology, Caroline A. Jones, ed. *Sensorium. Embodied Experience, Technology, and Contemporary Art* (Cambridge, MA: The List Visual Arts Center, MIT Press, 2006).

6   M.M. Taylor (Canada) ed., Research and Technology Organisation of NATO Technical Report 30, 'Visualisation of Massive Military Datasets: Human Factors, Applications, and Technologies' (© RTO/NATO, 2001), xiii.

7   Taylor, Research and Technology Organisation of NATO Technical Report 30.

tradition of hermeneutics, supports the benefits of an inclusive approach to understanding how we use Google Earth. Media analyst Chris Russill suggests such an approach with his idea of 'dual use':

> an industrial policy that acknowledges the formative influence of military design and funding on technology... while suggesting that civilian and peaceful applications are just as likely as not.[8]

This concept of dual use well describes the double-edged sword of remote sensing satellites and related digital mapping technologies. There is a particular kind of online group that uses these technologies, and it is perhaps more accurate to call this an online grouping because it consists of so many individual groups. This grouping shares one specific task and can be named in the broad sense as 'the military' – a term denoting a wide range of services, people, politics, economies, nation states, weaponry and other technologies. The broad task of the military can be simply described as actions and planning towards the waging of war or pre-emptive activities designed to defend and/ or expand the interests of the nation state or other entity to which the military in question belongs. The specific militaries that use satellite technologies are usually assumed to be those that have satellite launch capabilities and/or at least own their own communication satellites. In the Western academic domains of social science, media studies and textual analysis, the actions of the US military are those that are most often critiqued, although such critique is still in its early stages. Military groups, however, are not constrained by the boundaries imposed by nation states, as is clear from the continuing conflicts in the Middle East. Google Earth maps as well as hacks into other mapping programmes might well provide smaller overt or covert military organizations with tools to engage in warfare. And of course, they can use Google Earth in the same way that activist organizations use it, as explored in Chapter 4.

Google Earth also inadvertently enters disputes about state borders via its Global Awareness Layer 'Borders and Labels'. For example, on 20 July 2006, the unofficial Google Earth Blog posted an article about how a member of the Google Earth Community found a large-scale model in China via

---

[8]   Russill, 'Earth Observing Media', 280.

Google Earth and tracked it down as a representation of a disputed hilltop area between China and India, claimed by India at the time. If you follow the links in this posting and open the tagged information on Google Earth, you can easily follow how this 'discovery' was made: http://www.gearthblog. com/blog/archives/2006/07/huge_scale_mode.html.[9] This case is a good example of how Google Earth is being used by sometimes expert but often non-expert members of the general public for analysing and speculating on military objects, personnel and military matters in general. In a sense the public is using Google Earth for accessing what the public understands to be surveillance data and is using it to make speculative analyses of what it understands to be political and military information that is grounded in the actual world.[10] It is worth noting too that even if an analysis is made by an amateur, it does not mean that it is necessarily wrong. It is possible to see this kind of amateur usage of Google Earth to be another example of how the platform opens up sophisticated, complex means of surveillance to the general public; or at least, it opens up the ideas about what kinds of analysis can be and is being done by military intelligence communities on data collected via remote sensing satellites.

Google Earth's visualization of data via remote sensing satellites necessarily connotes the surveillance of geographic space over time. The platform thereby attracts comparison with surveillance for political and overtly militaristic purposes. You could say it is 'tainted' with the inherent malevolence of what militaries do in times of war. And it is interesting that the name of the company Keyhole recalls the US Keyhole satellite intelligence programme, which was instigated in the Cold War of the 1950s. This programme used 'KH reconnaissance satellites', otherwise known as spy satellites.[11] The information derived from these satellites was in the form of

---

[9] Google Earth Blog, http://www.gearthblog.com/blog/archives/2006/07/huge_scale_mode.html (accessed 18 December 2013).

[10] For another example of the huge amount of so-called intelligence analysis by the general public, see also Noah Schachtman, 'What Did Google Earth Spot in the Chinese Desert? Even an ex-CIA Analyst Isn't Sure', *Wired*, 1 September 2013, http://www.wired.com/dangerroom/2013/01/google-earth-china-hunh/#slideid-130941 (accessed 18 December 2013).

[11] There is a considerable amount of literature on the US Keyhole project, easily searchable on Google. See, for example, Pat Norris, *Watching Earth from Space: How Surveillance Helps Us – and Harms Us* (Dordrecht: Springer, 2010).

footage taken by analogue camera technologies. As noted earlier, Google bought Keyhole in 2004 and developed Keyhole's Earth Viewer spatial data visualization application suite into what became Google Earth in 2005. In a broad sense then, Google Earth uses technologies of remote sensing in many ways and for many reasons that can improve people's circumstances or bring them harm, sometimes simultaneously. Google Earth, as a massively accessible programme, can be used by anyone for any purpose, with the proviso that its usage is allowed by governments who monitor directly or indirectly their citizens' Internet activities.

Digital mapping

As noted in previous chapters, one of the illusions of Google Earth is that it is a static site, that the information it offers has an indexical relationship to time and space. This illusion that the status of the geographical information is beyond rational argument is common to all mapping enterprises. Google Earth's holistic aesthetic, as exemplified by the blue globe of Earth/Gaia, suggests to our perceptual gaze an even stronger connection between seeing and believing the information available in this site. In Google Earth, however, we only see those satellite images Google makes available at any particular time. So the images might not be the same as those we saw on a previous viewing, and these changes are not announced in the site itself apart from the time and space coordinates noted in very small print on the bottom of the images.

There is now a Google Earth Layer released by AGI Analytics in 2008, which shows the position of all known satellites and refreshes their position to match real-time coordinates. A user can therefore map changes in viewing positions of remote sensing satellites at the same time as she views a particular image of Earth.[12] A user can see via this layer the satellite most likely to have contributed data for that image. It makes clear that although we can return to a particular place on earth, we are not returning necessarily to the same

---

[12] See Parks, 'Earth Observation and Signal Territories', 298.

time/space image, even if it does seem that we are looking at the Earth as a static ever-present space.

Jason Farman's 2010 critique of Google Earth in the context of cartography well describes the implications for embodied responses to new media practices that arise from engaging with 'a new form of spatial interaction'.[13] This new kind of understanding about where we are spatially emerges from the nexus of human responses to digital, yet indexically linked, representations of space. His analysis focuses on Google's creation of the Google Earth Community (via the Bulletin Board System) that allows a community of users to interact with each other, to manipulate and mark/tag data from Google Earth as they illustrate and engage their own individual agendas as well as those of a broader sociopolitical nature. Another inherent link between the user of the application and its makers needs to be kept in mind: that of the consumer to a promoted product. This particular aspect must be noted, even if briefly, because Google Earth invites people to use it for commercial purposes, just as it offers grants to non-profit organizations to develop sites within its Google Earth Outreach programme.

For all its beauty, egalitarianism and military prehistory, Google Earth is and needs to be a product intended for corporate profit. For instance, Google insists on its branding logo being included in any images copied from the site: very reasonable in terms of copyright but also a reminder that data from Google Earth is derived from corporate interests whose agenda must embrace the knowledge that, in Parks' words, 'Satellite image data only becomes a document of the "real" and an index of the "historical" if there is reason to suspect it has relevance to current affairs.'[14] For example, although Google Earth has been used to trace human rights violations in Burma via satellite images,[15] Google Earth's branding of this kind of data as its own quickly followed.[16] This pragmatic approach to making sure that users know who 'owns'/has copyright of the image is not unusual; Google Earth, however, becomes highly visible as a corporation that owns images. As

---

[13] Farman, 'Mapping the Digital Empire', 560885.

[14] Parks, *Cultures in Orbit*, 91.

[15] See Crampton, 'Cartography: Maps 2.0', 93.

[16] For a critique on this aspect of 'branded information', see Lisa Parks, 'Digging into Google Earth', 91.

the most publicly accessible GIS platform and one that draws on public input via layer creation, there is perhaps a public perception in developed (usually financially privileged) countries that since we all live on Earth, all images of Earth from space should belong to everyone. Resentment can emerge as a public response to such a site – a resentment that exists alongside a sense of relief that this imaging of Earth is nevertheless available without too much technical or financial difficulty.

Such responses constitute part of the phenomenon described by new media theorist Henry Jenkins in his book *Convergence Culture: Where Old and New Media Collide.*[17] Jenkins follows up his work on 'fandom' with research on how fandom has changed in the digital age. Previous to Web 2.0, fans wrote and acted out their devotion to actors, television series, books, video games, anime, comics and films in the actual worlds of letter writing and character 'cosplay'.[18] Blogs are now associated with most of the above but the interesting phenomenon that Jenkins notes is the way in which fans mark their involvement with a character or narrative by presenting their own versions online. This means that production companies, publishers and authors find that their products have a life of their own, not only in the imaginations of their audiences but also in the virtual world of the World Wide Web, a world with the same kind of consequences as the actual. Is money at stake? Is copyright breached? Is the new fan-written ending to television series better than the one broadcast? Jenkins addresses such questions and describes some of the interactions between this kind of online fandom and the copyright owners of the original product.

I suggest that the kinds of convergences that Jenkins describes in the context of fandom are applicable to the world of digital mapping. These convergences happen most significantly between old and new communication technologies when the new technologies are at the stage of replacing many uses of the old, or are offering new uses, especially when new uses are offered to people who have had access to the older technology only

---

[17] Henry Jenkins, *Convergence Culture: Where Old and New Media Collide* (New York, NY and London: New York University Press, 2006).

[18] For a current definition, see Melissa de Zwort, 'Cosplay, Creativity and Immaterial Labours of Love', in *Amateur Media. Social, Cultural and Legal Perspectives*, eds Dan Hunter et al. (New York, NY: Routledge, 2013), 171.

as a consumer, a recipient, as an audience – not as a maker. A wonderfully graphic example of how Google Earth map imagery is culturally convergent and not all that far away from the maps imagined and executed over the centuries can be found in Gurevitch's Figure 2. 'Venice as visualized by Braun and Hogenberg (1572) and Google Earth (2010)' in his article 'Digital Globe as Climatic Coming Attraction'.[19]

As noted in Chapter 1, Crampton (2009) draws attention to a specific attribute of the new kind of mapping which he calls the 'geoweb'. He is particularly interested in its use of 'citizen-orientated map-making efforts'.[20] In his words, 'Such [remote sensing satellite] imagery, alongside the tremendous possibilities of 'crowd sourced' geospatial data, represents interesting new developments in cartography'.[21] As Crampton describes, Web 2.0, 'the social web', invites the public to contribute to and communicate with each other via the Internet and via a vast range of both publicly and privately owned software programs: from personal blogs to Facebook. Digital mapping, and most particularly Google Earth, offers free access to huge mapping sites and invites the public to contribute information via tagging and layer creation. Once given, this information becomes part of the Google Earth suite of spatial data. As Parks (2013) notes, when even government-owned and -produced data such as that provided by the US Federal Communication Commission 'is accessed and viewed in Google Earth, it becomes the privately owned intellectual property of Google'.[22] Parks goes on to point out that in fact 'every layer, which is typically generated through the "free labor" of private individuals, non-profit organizations, government agencies, or companies, is not only an inscription in Google Earth, but also is an investment in and gift to Google ...'.[23]

This somewhat disturbing 'handing over' of information occurs of course not only in Google Earth but also in other corporately owned social media platforms such as Facebook. As long as the information that this is so is

[19] Leon Gurevitch, 'Digital Globe as Climatic Coming Attraction: From Theatrical Release to Theatre of War', *Canadian Journal of Communication* 38 (2013): 339.

[20] Crampton, 'Cartography: Maps 2.0', 91.

[21] Crampton, 'Cartography: Maps 2.0', 91.

[22] Parks, 'Earth Observation and Signal Territories', 297–298.

[23] Parks, 'Earth Observation and Signal Territories', 298.

easily accessible to users in various statements of 'terms and conditions of use' documents, then is it possible to talk about this particular *quid pro quo* as based on the concept of informed consent? I suggest that users are most often so bedazzled by the array of digital communication platforms that people might easily underestimate the risks of handing over intellectual property to Google and providing public access to their work or probably often, not even read the terms and conditions. Digital mapping now pervades wealthy societies through the use of mobile devices and particularly now through 'smart phones' and tablets. Farman says that we should no longer consider our mobile electronic computer devices as prostheses (in Haraway's sense) but that we need to examine how mobile devices (and I suggest we can include laptop computers here) are actually changing the status of how we are embodied in space: social space, virtual space and actual space. Farman claims that 'The very practice of embodied space is becoming entirely reliant on the seamless interaction between our devices and our landscapes'[24] and that representations of space are now 'not outside of the lived experience of that space'.[25] He recalls Baudrillard's maxim: 'The territory no longer precedes the map, nor does it survive it'.[26] In other words, his suggestion here is that with the advent of interactive digital mapping, our sense of where our body is in space and time now depends on how it is represented to us on mobile computer devices. The potential for Orwellian surveillance and other deceptive scenarios is obvious. So it becomes interesting when a huge corporation like Google decides to embark on outreach programmes that it says are designed to show the public how Google welcomes the role of a socially responsible global corporate citizen.

Outreach

It is hopefully by now a truism that the consequences of a particular technology can extend way beyond the intentions or imaginations of its

---

[24] Farman, *Mobile Interface Theory*, 46.

[25] Farman, *Mobile Interface Theory*, 46.

[26] Jean Baudrillard, *Simulacra and Simulation*, 1, as quoted in Jason Farman, *Mobile Interface Theory*, 46.

inventors and that its applications can also reach far outside of a user's primary intention. In the context of photography for example, science photographer Felice Frankel comments as follows:

> I have recently become aware that the visual impact itself of the photographs I make in the lab can have significant consequences, allowing them to communicate important information about science research not only to other scientists in the lab, or in the field, but to a broader, non-scientific public as well.[27]

Frankel's words recall Virilio's conceptualization of 'the accident' in the forever present of current society's imagination of self.[28] Virilio points out the fallacy in believing that accidents are born only of chance; rather he locates the 'catastrophes' that are described by media of various kinds as accidents, as inherent to the technologies through which they happen and/ or inherent to the environment of Earth and humanity's interactions with this environment. In other words, 'accidents happen', they are not unexpected *per se*.

Institutions invariably and, in their own terms of operation, necessarily try to manipulate how their various access portals and products are used. One of the reasons for this is clearly to capitalize as much as possible on their commodities, but I think we can say that another perhaps less acknowledged reason is to present themselves as socially responsible and so avoid or be seen to avoid some of the more negative outcomes that can emerge from the use of their commodities. One way to do this is to attach the word 'outreach' to a suite of products that are branded and/or otherwise carefully associated with or sponsored by the institution. The current and common understanding of the term 'outreach' is in the context of an organization communicating to people outside its own internal parameters of operation. Although the term connotes

---

[27] Felice Frankel, *Science*, 280, 1698–1700, 12 June 1998, quoted in the Preface of The Research and Technology Organisation (RTO) of NATO RTO Technical Report 30, 2001 © RTO/NATO. ISBN 92-837-1066-5. 'Visualisation of Massive Military Datasets: Human Factors, Applications, and Technologies'. This Technical Report represents the Final Report of IST-013/RTG-002 submitted by the members of IST-013/RTG-002 for the RTO Information Systems Technology Panel (IST) M.M. Taylor (Canada), eds J.G. Hollands et al. (2001).

[28] For Virilio's argument on 'The Invention of Accidents', see Paul Virilio, *The Original Accident*, trans. Julie Rose (Cambridge: Polity Press, 2005), 9–14.

a gesture of 'reaching out to help', currently it firmly denotes those offerings directed towards the social good that an organization might make towards other organizations, less formal groupings of people and sometimes even to people as individuals.

To include the term 'outreach' in any act of self-definition is to imply a generalized gesture of generosity, of wanting to contribute somehow to the social good. Whether or not claiming an attitude of generosity towards the public is itself disingenuous, is an issue to be considered, although to some extent any discussion of motives and outcomes relies on published agendas and analysis of their consequences. This issue is taken up explicitly in Chapter 5, where I consider the discursive place which is my journeying into the experience of Google Earth that my writing of this book has given to me. In the next chapter, I address this attitude of generosity as it happens in the context of activism via Google Earth and how effective it is in a specific case study. First, however, I think it is necessary to focus on some of the actual user groups and what they want to use Google Earth for.

I want to make a small proviso here. During all the discussion in this chapter, there is an inherent danger of falling into a severe case of 'google speak'. This is because I am describing tools developed for using Google Earth, why Google has developed them and because I am also quoting Google.com web pages as part of these descriptions. Whilst it is naïve to claim that Google's descriptions of its own operations and products are transparently the 'truth' (even if this is Google's intention), I think it is nevertheless useful to include these descriptions as ones that are necessary for the process of deciding why we use Google Earth and associated products.

Google Earth Outreach

Google Earth Outreach is a highly purposeful set of applications.[29] The programme was founded by Rebecca Moore, a professional software

---

[29] See 'Google Outreach Home', http://www.google.com/earth/outreach/index.html (accessed 3 January 2014).

engineer who worked for Google Earth, and now leads Google Earth Outreach and also the Google Earth Engine programme. Google introduced Google Earth Outreach in June 2007. This programme was a direct result of Moore's innovative use in 2005 of Google Earth (launched only two months earlier) on behalf of a community action group called 'Neighbors Against Irresponsible Logging'. A water company planned to log a six mile swathe of over 1,000 acres of redwood forest. The logging plan map was 'a black and white, low-resolution sketch that did not convey what was at stake … local citizens did not understand it'.[30] Moore mapped the area in Google Earth and added a layer that showed (in red) what areas in the environment and around the community itself would be affected. She then created a flyover animation that was shown to politicians. The flyover gave easy access to information that was otherwise contained/obscured in a 400-page document.[31] The mapping and the flyover were used subsequently in a legal process to prove that the logging plan itself was actually illegal. The logging company claimed that Google Earth was not a proper information tool, that it was just a toy; Moore was able to respond that the information came from objective satellite-sourced visualization data and in short, the plan did not go ahead.[32] As a result of this successful environmental campaign, Google Earth gained a profile as a conduit for environmental and other kinds of social justice advocacy. Moore then was able to launch Google Earth Outreach as a 20 per cent project: a project that a Google employer was passionate about and could spend one day a week working on. Moore became Google Earth's liaison for 'grass roots' community issues.

In September 2005, Hurricane Katrina landed on New Orleans with terrible consequences both during the hurricane and in its aftermath. The US Government sent planes to photograph the overall devastation and specific areas where the levy banks had burst, but had no means of quickly publishing this visual information so that it could be used by first responders to the disaster. The Government asked Google whether they would publish this material in Google Earth and continually refresh it so

---

[30] Rebecca Moore, 'Raising Global Awareness with Google Earth. What Do You Do After Flying to Your Home?', *Imaging Notes* 22, no. 2 (2007): 6.

[31] GEOE Interview, 14 February 2014.

[32] GEOE Interview, 14 February 2014.

that emergency workers who came from all over the country and did not necessarily know the geography of the city, could find people calling for help, some of whom were stranded on the roofs of their houses by the floods. Google agreed and Google Earth employees worked 'around the clock' to provide a 'parallel Google Earth' that they constantly updated. These first responders told Google that they used Google Earth to save more than 4,000 lives.[33]

After Hurricane Katrina, NFP organizations began to approach Google Earth to find out whether there were ways in which Google Earth might be able to help them in their advocacy programmes. They asked the founder of Google Earth Outreach Rebecca Moore:

> Can Google Earth help us illustrate our projects in a new and more compelling manner than text and slideshows? Are there methods or tools for importing our existing data into Google Earth? Can you tell us about any other non-profits who've been successful at using GE to reach a new audience, raise awareness, gain volunteers, inspire people into action, and create a tangible impact?[34]

In response to these questions, the Google Earth Outreach programme was developed over the following year and was officially launched in June 2007. The Outreach development team consisted of programmers very interested in the idea of putting information together to create a story.[35] Again, this is a significant point for understanding the role of Google Earth Outreach: its aim was and is to enable activist and advocacy groups to create a narrative that is easy for the non-specialist, the non-hacker and the non-cartographer to understand. The Google Earth Outreach website itself currently offers a showcase of materials (videos, photographs and text) that have been included as recent Google Earth Awareness Layers, with information about how to obtain funding from Google Earth to make these overlays, together with how to make them. Google Earth Layers can be updated by their creators.

[33] GEOE Interview, 14 February 2014.

[34] Rebecca Moore, 'Introducing Google Earth Outreach', Google Official Blog, 26 June 2007. http:// googleblog.blogspot.com.au/2007/06/introducing-google-earth-outreach.html (accessed 31 March 2014).

[35] GEOE Interview, 14 February 2014.

Google Earth Outreach has entered into several collaborations with environmental groups. One major outcome is the United Nations Environment Programme (UNEP) layer: Atlas of Our Changing World. This atlas had been published in hard copy in 2005; in September 2006 (before the official launch of the Outreach programme) Google and UNEP jointly published the online version as a layer in Google Earth.[36] The role of Google Earth Outreach in providing information and technology 'to make the world a better place' has been recognized recently by UNEP via its Champions of the Earth Laureate award in 2013 to Brian McClendon, Co-founder and Vice-President of Google Earth, for Entrepreneurial Vision, United States.[37] In Chapter 4, I investigate an important collaboration: an example of humanitarian efforts achieved through Google Earth Outreach. This was a collaboration between Google Earth and the United States Holocaust Memorial Museum (USHMM): a layer called 'Crisis in Darfur', overlaid on top of the Google Earth image of Northern Africa.

Other major collaborations are listed as Global Awareness Layers in the first page that opens in Google Earth. These include overlays from Greenpeace, UNICEF Water and Sanitation, Fair Trade Certified, WWF Conservation Projects and USHMM: World is Witness.

A project that Moore has been personally involved with is 'The Elders: Every Human Has Rights' overlay. This project began with an approach in 2007 to Google Earth Outreach by Brazilian Amazon tribal elder Almir Surui. He told Outreach about his efforts to halt illegal logging in his tribe's home area in the Amazon rainforest, and asked Outreach to teach the tribe how to bring the logging issue into the political and legal arena, saying that it was 'time to put down the bow and arrow and take up the laptop'.[38] Surui said that the use of technology like Google Earth formed a bridge between young people in the tribe who enjoyed using the digital technology, and elders who could see their homelands clearly depicted. He told my Google Earth

---

[36] Moore, 'Raising Global Awareness with Google Earth', 4.
[37] See UNEP News Centre, 'Brian McClendon, co-founder and VP of Google Earth Awarded Top UN Environment Prize for Mapping New Conservation Paths and Creating Livelihood Opportunities through the Green Economy'. http://www.unep.org/NewsCentre/default.aspx?DocumentID=2726&ArticleID=9621 (accessed 31 March 2014).
[38] GEOE Interview, 14 February 2014.

Outreach interviewee that Google Earth is a particularly indigenous way for telling a story because it ties the land with issues and dreams.[39]

In Australia, Moore has also been directly instrumental in assisting with publishing KML files for the project: Ngarluma Ngurra: Aboriginal Culture on the Map.[40] This culture-mapping project is funded by the Western Australian organization, FORM – Building a State of Creativity. The digital publication that was produced in conjunction with this project includes a section by Moore that clearly articulates the aims and outcomes of Google Earth Outreach collaborations with indigenous communities.[41] The following Google Earth Outreach web page gives a list of other stories and collaborations: http://issuu.com/form-wa/docs/ngarluma_ngurra_catalogue_4ec7aebf3b297f (31 March 2014).

Google has been moving towards a 'unified web approach' to information generated in Google Earth and has been working towards providing a new service with Google indexing software to provide a system that would allow people to discover layers and maps more easily, that would increase 'discoverability' on Google Earth.[42] On 28 February 2014, the Google Earth Official Blog announced the launch of the 'Google Maps Gallery' with the article 'Google Releases the Google Maps Gallery, with solid support for Google Earth'.[43] Google Maps Gallery[44] allows the user to search for maps and to load their own maps. It also showcases maps that Google considers are well made and of general interest in their 'top maps' displays. Makers of these latter layers will be able to write a blog about the layer or Google Earth will write it for them. Makers can refresh layers as often as they want. For example, the US Geological Survey Office has an Earthquakes Layer that is refreshed every 4–5 minutes.[45] The intention behind the creation of the Google Maps Gallery

[39] GEOE Interview, 14 February 2014.
[40] For a link to 'Ngarluma Ngurra Google Earth Tour', see http://www.form.net.au/project/ngarluma-ngurra/ (accessed 31 March 2014).
[41] For a link to this catalogue publication, *Ngarluma Ngurra: Aboriginal Culture on the map*, Catalogue, 2012, http://issuu.com/form-wa/docs/ngarluma_ngurra_catalogue_4ec7aebf3b297f (accessed 31 March 2014). Rebecca Moore's essay 'Google Earth Outreach: Seeing Is Believing' is on pages 19–20.
[42] GEOE Interview, 14 February 2014.
[43] Google Earth Blog, http://www.gearthblog.com/blog/archives/2014/02/google-releases-google-maps-gallery-solid-support-google-earth.html (accessed 31 March 2014).
[44] Google Maps Gallery, http://maps.google.com/gallery/
[45] GEOE Interview, 14 February 2014.

is that people can use 2D or 3D maps provided by Google Maps and Google Earth to create layers, which can then be uploaded and searched for.

In December 2004, however, Google had already introduced two sets of technologies specifically 'dedicated to the idea that technology can help make the world a better place'.[46] They called these Google.org and the Google Earth Engine. To quote Google.com:

> Google.org develops technologies to help address global challenges and supports innovative partners through grants, investments and in-kind resources
>
> … we've developed a broad range of beneficial technologies. Highlights include creating Google Person Finder, which helps reconnect people in the wake of major disasters; developing Flu and Dengue Trends, which use search trends to provide early warning systems for possible disease outbreaks …[47]

Google.org also delivers Google Crisis Response.[48] This application collates and provides information in emergency situations, including 'public alerts' about the developing emergence of a crisis and information to the public about shelters and other assistance. It also provides tools so that first responders can live stream information about a crisis to each other. The home page for this application currently links to examples of how it has been used in the past and links descriptions of tools that have been developed for first responders, including the Google Crisis Map. Google Earth can also be used for viewing before and after views of a disaster site, and for creating and editing layers containing information about a crisis.

Google Earth Engine was built 'to enable scientists, governments and native tribes to monitor changes to the Earth's surface'.[49] It is not open for use and development by everyone: in Google's words, 'Access to Earth Engine is currently available as a limited release to a small group of partners. If you are interested in developing on the Earth Engine platform, let us know.'[50]

[46] About Google.org, http://www.google.org/about.html (accessed 4 December 2013).
[47] Google Earth Engine, https://earthengine.google.org/–intro (accessed 4 December 2013).
[48] Google Crisis Response, http://www.google.org/crisisresponse/ (accessed 26 December 2014).
[49] Google Crisis Response, http://www.google.org/crisisresponse/.
[50] Google Crisis Response, http://www.google.org/crisisresponse/. Also see Google Earth Engine, https://earthengine.google.org/-intro (accessed 4 December 2013).

On the introductory web page, however, there is a gallery of Landsat time-lapse videos (1984–2012) of several locations and geographies that show what kind of visual data and analyses can be derived using the Google Earth Engine.

Google.com offers several products associated with Google Earth that allow a more sophisticated use of Google's 3D digital representations of Earth. Broadly speaking, people can use Google Earth in its 'free' stand-alone version, or through applying to use the Google Earth Engine, or through the advanced functions offered in the Google Earth Pro programme. All these products come under the rubric of Google Earth Outreach. Although individuals can and do use Google Earth Pro for personal and private reasons, primary users are organizations and institutions: business companies and corporations, NFP organizations, non-government organizations (NGOs), government departments and academic or industry-based researchers. As is the case with most categorizing processes, these categories often overlap with each other.

According to their specific tasks, these groups want to use Google Earth as a tool for mapping, for analysing online data, especially geospatial data, and for representing/visualizing that data. Google.com invites these groups to buy an advanced set of functions provided as Google Earth Pro. According to Google.com, Google Earth Pro offers the following properties:

- Advanced measurements: Measure parking lots and land developments with polygon area measure or determine affected radius with circle measure.
- High-resolution printing: Print images up to 4,800 × 3,200 pixel resolution.
- Exclusive Pro data layers: Demographics, parcels and traffic count.
- Spreadsheet import: Ingest up to 2,500 addresses at a time, assigning placemarks and style templates in bulk.
- GIS import: Visualize ESRI shapefiles (.shp) and MapInfo (.tab) files.
- Movie-Maker: Export Windows Media and Quicktime HD movies, up to 1,920 × 1,080 resolution.

- Support: Email assistance with downloading, installation, activation and account management.[51]

Google Earth Pro now requires the payment of a fee, although free trials are available and the fee is not huge. In this sense, like Google Earth Engine, it is a more exclusive way to extend the usage of Google Earth's stand-alone function as a provider of images of Earth via remote sensing satellite technology. Google web pages comprehensively list how Google Earth Pro works, and support is included in the annual fee, as noted above. Google promotes this set of applications as useful for 'professional' users. At the time of writing, Google is using Rockware, Colorado as an example of a business that is using Google Earth Pro to provide above-ground mapping to their core business of underground mapping.[52]

Using Google Earth

Why do governments use Google Earth instead of virtual globes such as NASA's World Wind and ESRI's ArcGIS Explorer? The simple answer to this question at this moment in time is: because it is the platform most easily available to and most often used by a public who is consequently more accustomed now to Google Earth's particular iconography, interactive tools and overall aesthetic than any other virtual globe. The use of Google Earth as an interface between geospatial data and people who are not highly literate in GIS technologies can itself be understood as a liberal action, in that people who could not receive remote sensing data before, now can.

Similarly, Google has made available free to the public the facility Google Tables through which already published or otherwise available Fusion Tables[53] can be merged with the user's own data set and the result embedded

---

[51] I asked the question 'What Is Google Earth Pro?' This support page replied – on https://support. google.com/earth/answer/189188?hl=en (accessed 7 December 2013).

[52] See http://www.google.com/enterprise/mapsearth/products/earthpro.html (accessed 9 December 2013).

[53] Google Support, 'About Fusion Tables: Fusion Tables is an experimental data visualisation web application to gather, visualize, and share larger data tables.' https://support.google.com/ fusiontables/answer/2571232?hl=en&ref_topic=1652595 (accessed 6 April 2014).

into Google sites. Google provides online tutorials via YouTube to teach people how to create and annotate their own maps.[54] Google also offers businesses, researchers and government institutions the ability to build their own version of Google Earth with the Google Earth Builder. This product offers facilities to organizations for storing and processing their GIS data on the Google's enormous server farms. The product then allows this information to be displayed in Google Maps, Google Earth or on Android phones.[55]

Google Earth Builder is one of four products that Google designed for use by professional organizations, the others being Google Maps API for Business, Google Earth Pro and Google Earth Enterprise.[56] Matteo Luccio, who writes in and edits magazines on GIS, says that people often confuse the Google Earth software that produces the images with the images themselves. They are separate and Google Earth owns imagery which is not necessarily also licensed by people buying the Google Earth Builder product. Organizations can create their own virtual globes this way, upload their own data to the Google cloud and have control over who sees the imagery and thereby who has access to the data they upload to the Google cloud. It is easy to argue with Luccio that by developing the Google Earth Builder, Google is basically commodifying its storage space on its server farms and that 'Google does not want to be seen as [only] a provider of imagery, but [also a provider] of the capability to manage it.'[57]

At this point in my discussion, however, there is an even greater danger of appearing to slide into a discourse that sounds very like Google's own brand of online advertising, but the discussion is necessary if also speculative. This is because there needs to be a way to articulate why there is still a Google Earth presence in situations that can be met perhaps technically even better by other platforms. Here are examples of how and

[54] Google Support, 'Create with Fusion Tables', https://support.google.com/fusiontables/answer/184641?hl=en (accessed 6 April 2014).

[55] Luccio, Matteo, 'Google Earth Builder. Productizing Server Farms for Storing and Processing Geospatial Data', *Imaging Notes* 27, no. 2 (2012), http://www.imagingnotes.com/go/article_free].php?mp_id=303 (accessed 6 April 2014).

[56] Luccio, *Imaging Notes*.

[57] Luccio, *Imaging Notes*.

why Google Earth might be incorporated in complex viewing layers on sites that act as portals for communication between government bodies and the communities they serve.

The Territory and Municipal Services (TAMS) is a broad grouping within the Australian Capital Territory Government. The services they offer include those under the titles of Parks and Recreation (including fire management of wilderness areas within the Territory), Libraries, Pets, Recycling and Waste, Recreational Activities and the Government's portal for information on Roads and Traffic. The people who work in TAMS need to analyse information and format it in ways that are useful to colleagues in their own area and for colleagues in other areas. When such collated information is then presented to the public, the formatting of information becomes crucial, and given that the information is for residents of a city (Canberra, Australia's capital city), then much of this information is provided in the form of a map. Increasingly, such maps are being created as interactive, so that information about a particular location can be accessed on mobile digital technologies (phones and tablets) via tagged window layers on a digital map or even via scan sites at the location itself. With its aerial mapping and analytic GIS products, ESRI[58] is the primary programme used by TAMS.

On World GIS Day 2013 (20 November) I attended seminars organized by TAMS that offered some insight into what was needed for their tasks. The morning seminar was given by people from ESRI, promoting one of their newest products: ArcGIS Online, 'a cloud-based mapping system for organizations that offers collaboration tools for cataloging, visualizing, and sharing geospatial information.'[59] It became clear that ESRI was TAMS' product of choice, although why it is so is not yet quite clear to me, except that the department was accustomed to using a combination of LiDAR ESRI imagery and Google Maps, and was comfortable with this, possibly as a legacy technology.

The afternoon seminar was given by the ICT Manager for the Fire Division in the Department of Environment and Primary Industries of the

---

[58] ESRI, www.esri.com
[59] ESRI, www.esri.com

State Government of Victoria, Australia. The Fire Division had been tasked by the 2009 Victorian Government Bushfires Royal Commission[60] to find a 'new paradigm' for founding a new kind of 'interoperability' between groups of people who were significant in acting to prevent and fight bushfires in Victoria. The outcome has been the creation of the Fire Web portal that was designed to allow live streaming of mapping information to operational online groups. The logistics of this enterprise are obviously very complex since the information layers are so dense, with more than 300 data sets. The Fire Division uses ESRI for spatial data processing and web-based map making. The eventual aim is to allow the general public in relevant threatened communities to have access to information that predicts the possible spread pattern of a fire. There is a KML feed for Google Earth on the dashboard of the Emergency Map, however, and Google Earth is recognized by the organization as the most easily accessible and most often accessed viewing layer by the general public, by both residents and community-based firefighters. In other words, Google Earth is most likely, the most accessible way to allow the general public to access emergency fire information.

Here is another Australian example of a government department using Google Earth. The Australian Commonwealth Department of the Environment Sustainability Policy and Analysis Division has an Environmental Resources Information Network (ERIN) branch.[61] Approximately forty GIS specialists work within ERIN, whose overall function within the department is to find, manage, deploy and distribute environmental information primarily to colleagues within the department, and with other state and commonwealth agencies, for example, Geoscience Australia and the Commonwealth Scientific and Industrial Research Organisation. To quote an ERIN employee who works in the Park and Marine Team, the various teams assist intradepartmental groups

---

[60] '2009 Victorian Bushfires Royal Commission Final Report', http://www.royalcommission.vic.gov. au/commission-reports/final-report (accessed 6 December 2013). This Royal Commission Inquiry was instigated to address the catastrophic fires in the Australian State of Victoria that peaked on 'Black Saturday', 7 February 2009. A total of 173 people died as a result of these fires. The radar map coding of cloud activity over the burning areas on that day showed as if there was thunderstorm activity; the radar was actually showing the dense clouds, the pyrocumulus clouds created by the firestorms themselves.

[61] For a look at the wide range of departmental functions supported by ERIN, see http://www. environment.gov.au/aggregation/topics (accessed 9 January 2014).

with their spatial information and analysis requirements. Mostly this involves making maps, although we do some research in such areas as 'connectivity' in the landscape (least cost pathways). We're also starting to provide quite a bit of support for online mapping and report generation.[62]

He goes on to say that 'We also deal with State-based environmental information agencies (as they can often provide us with information that is then collated into a national data set) as well as Museums, Herbaria, Indigenous Management Groups, NGOs, academics/researchers and occasionally Local Govt.'[63]

When I asked why his area used Google Earth, he answered as follows:

It's an easy tool to use; has the ability to incorporate different spatial data file types (the ones we would use the most would be ESRI.shp files and .gpx files); has good (high resolution) imagery for a lot of the areas that we're interested in; the imagery is 'free' and the Pro version annual costs are low; our client/policy areas of the Dept are used to and often request we use satellite image backdrops; Non-technical users are getting more and more used to applications like Google Maps and Google Earth; we use GE Pro to help verify information (usually in the form of a.jpg image or hard copy map) we receive from our line areas or external clients we deal with.[64]

At the time of writing, ERIN holds fifteen Google Pro licences which are renewed annually. The Google Pro application is also preferred 'as it does allow us to "cut out" pieces of imagery that we can then use in mapping exercises or provide to our client for use in handheld GPS's, like Garmins.'[65]

The NSW State Government's Department of Finance and Services' introduction of the NSW Globe[66] in late 2013 provides a new example of how GIS-associated information is being offered to the public by a government body, via Google Earth. The NSW Globe offers Land Property Information online and for free. It is an open data set that uses Google Earth for

---

[62] Email interview, 7 January 2014.
[63] Email interview, 7 January 2014.
[64] Email interview, 7 January 2014.
[65] Email interview, 7 January 2014.
[66] NSW Land and Property Information, NSW Department of Finance and Services, 'Explore the NSW Globe', http://globe.six.nsw.gov.au/ (accessed 6 April 2014).

visualization of geospatial information that includes rail and road networks and property boundaries and historical image data on floods, fires and other emergencies.[67] Clearly then, Google Earth is being used by the Australian Federal and State Governments at several levels as a way of constructing, analysing and distributing spatial data and imaging across a wide range of people with varying degrees of digital GIS literacy. Drawing on the above examples and the fact that Google Earth is free to download, I suggest that it is possible to say now that Google Earth is often the first point of call for creating an information interface between GIS specialists and the larger community.

Using Google Earth for research and education

In late 2013, I did a search in Google Scholar and linked sites for research and journal articles that significantly referred to Google Earth. My search outcome was a snapshot of who uses Google Earth for research, why and when up until 2013. The range of topics is extensive. Around the time of Google Earth's launch, the whole phenomenon of virtual globes itself was still being researched. In the first few years after the launch, researchers predictably were looking more at the possibilities for using Google Earth for data collection, mapping and how to include these functions for presentation and publication. Here are some examples. The UK NGO MapAction presented the 2008 paper 'Google Earth and its potential in the humanitarian sector: a briefing paper.'[68] In 2007, Michael F. Goodchild from the National Center for Geographic Information and Analysis, University of California, published 'Citizen as sensors: the world of volunteered geography.'[69] This paper compared earlier sites such as Wikimapia and OpenStreetMap with Google Earth, focusing on the digital mapping phenomenon of citizen volunteering

[67] Stephen Woodhouse, Chief Information and Technology Officer, NSW Land and Property Information, 'NSW Globe', http://finance.nsw.gov.au/ict/sites/default/files/4.%20NSW%20Globe.pdf (accessed 6 April 2014).

[68] MapAction, 'Google Earth and its potential in the humanitarian sector: a briefing paper' (2008). http://www.mapaction.org.

[69] Michael F. Goodchild, 'Citizens as Sensors: the World of Volunteered Geography', *GeoJournal* 69 (2007): 211–221. doi: 10.1007/s10708-007-9111-y.

information to digital mapping platforms. Authors from the Department of Biomedical Engineering, Linköpings Universitet, Sweden, wrote a conference paper in 2007 titled 'Graphical Overview and Navigation of Electronic Health Records in a prototyping environment using Google Earth and openEHR Archetypes'.[70]

From 2009, this trend of researching Google Earth as a valid component of research methodologies continued. For example, the 2009 paper 'Virtual globes and geospatial health: the potential of new tools in management and control of vector-borne diseases'[71] was written by biologists and epidemiologists in the areas of tropical, veterinary and community medicine. Another example from this period is the 2009 paper by geographers Michael Crutcher and Matthew Zook 'Placemarks and waterlines: Racialized cyberscapes in post-Katrina Google Earth'.[72] This paper provides a social critique of Google Earth's first outreach contribution that mapped damage in New Orleans after and during Hurricane Katrina. Google Earth is also being used as a teaching tool in geography teaching at both secondary and tertiary levels, as described and discussed in the Geological Society of America's Special Paper: 'Introduction: The Application of Google Geo Tools to geoscience and education research'.[73]

The trend of using Google Earth as a potential tool rather than just a phenomenon to be examined for itself continued with papers by social researchers such as Philippa Clarke from the Institute for Social Research, University of Michigan: 'Using Google Earth to conduct a neighborhood audit: Reliability of a virtual audit instrument'.[74] By 2010, Google Earth usage by researchers was moving towards being accepted as a valid

[70] Erik Sundvall et al., 'Graphical Overview and Navigation of Electronic Health Records in a Prototyping Environment Using Google Earth and openEHR Archetypes' in *MEDINFO 2007*, eds K. Kuhn et al. (Amsterdam: IOS Press, 2007): 1043–1047.

[71] Anna-Sofie Stensgaard et al., 'Virtual Globes and Geospatial Health: The Potential of New Tools in the Management and Control of Vector-Borne Diseases', *Geospatial Health* 3, no. 2 (2009): 127–141. www.GnosisGIS.org.

[72] Michael Crutcher, Matthew Zook, 'Placemarks and Waterlines: Racialized Cyberscapes in Post-Katrina Google Earth', *Geoforum* 40 (2009): 523–534. doi: 10.1016/j.geoforum.2009.01.003.

[73] J.E. Bailey, S.J. Whitmeyer and D.G. De Paor, 'Introduction: The Application of Google Geo Tools to Geoscience Education and Research', *Geological Society of America Special Papers* 492 (2012): vii–xix. doi:10.1130/2012.2492(00).

[74] Philippa Clarke et al., 'Using Google Earth to Conduct a Neighborhood Audit: Reliability of a Virtual Audit Instrument', *Health and Place* 16 (2010): 1224–1229. doi:10.1016/j.healthplace.2010.08.007.

and readily accessible research tool in several areas. An example from 2010 is by authors from the Institute for Research on Earth Evolution, Japan Agency for Marine-Earth Science and Technology: 'Visualisation of geoscience data on Google Earth: Development of a data converter system for seismic tomographic models'.[75] In his 2010 article 'Camp Delta, Google Earth and the ethics of remote sensing in archaeology',[76] Adrian Myers explicitly interrogates the use of Google Earth as methodology in archeology, illustrating that the platform was being used to an extent that required an analysis for its ethical implications. By 2013, Google Earth and its various analytic products were firmly positioned as accepted tools for research amid an array of other GIS. However, at the same time, the platform was still being interrogated for its usefulness; for example, 'Web GIS in practice X: a Microsoft Kinect natural user interface for Google Earth navigation'[77] (2011) by Kamel Boulos et al. from the Faculty of Health, University of Plymouth, UK.

Google Earth is a paradox: it offers a tool for serious research but via a platform that is promoted to the non-researching, non-professional community as much as it is directed towards 'serious' endeavours of the professional, business and scientific communities. Its flight simulator application and interactive interface particularly suggest analogies if not direct comparisons with the 'frivolous' leisure activities of computer gaming.

A ludic interface: Playing Google Earth

The success of the playful mode of reading is in inverse proportion to the cultural scorn poured on it.

Sean Cubitt[78]

---

[75] Yasuko Yamagishi et al., 'Visualization of Geoscience Data on Google Earth: Development of a Data Converter System for Seismic Topographic Models', *Computers and Geosciences* 36 (2010): 373–382. ISSN 0098–3004.

[76] Adrian Myers, 'Camp Delta, Google Earth and the Ethics of Remote Sensing in Archaeology', *World Archeology* 42, no. 3 (2010): 455–467. doi: 10.1080/00438243.2010.498640.

[77] Kamel Boulos et al., 'Web GIS in Practice X: A Microsoft Kinect Natural User Interface for Google Earth Navigation', *International Journal of Health Geographics* 10, no. 45 (2011). doi 10.1186/1476-072X-10-45.

[78] Cubitt, *Digital Aesthetics*, 15.

How much can our usage of Google Earth be thought of as 'playing'? Google Earth is a site for research, for playful fun, for getting information about the world that Google and remote sensing technologies see fit to provide. With its interactive, responsive interface, I agree with Munster that

> Perhaps, then, Google Earth shares something with massively multiplayer online role-playing games (MMPORGS) in that it builds a Google-like space to simply hang around in and 'play', beyond the instrumentalism of search.[79]

There are many ways in which we use the word 'play'[80] and several theorists who have written on both the concept and practice of play, the most prominent being anthropologist Gregory Bateson. He distinguished between 'play' and 'non-play' as distinctive modes of activity.[81] These days we might apply an analogy of this distinction to the difference between the virtual and the actual. In other words, just as play activities have consequences in the world outside the arena of play, do activities that take place in the virtual space created by computers have consequences in the actual? Bateson also understood play in childhood to be an important form of learning how to communicate. A specific kind of play has been called 'serious play'. This type of playing has to do with learning in the speculative world of acting and thinking 'as if' something was true or actual, but this process has less to do with having fun and more to do with playing in order to get information through processes of simulation and modelling. The processes involved in this kind of play can be uncomfortable, disturbing and frustrating.[82] With these ideas of play in mind, it is interesting then to consider again Munster's and Stahl's insights on how entering into Google Earth can feel like computer gaming.

---

[79] Munster. *An Aesthesia of Networks*, 61.

[80] See, for example, Richard Schechner, Chapter 4 'Play' in *Performance Studies: An Introduction*, Third edition (New York, NY: Routledge, 2013), 89–122.

[81] See Gregory Bateson, 'A Theory of Play and Fantasy', in *Steps to an Ecology of Mind. Collected Essays in Anthropology, Psychiatry, Evolution and Epistemology (1972)* (Northvale, NJ: Jason Aronson Inc, 1987), 183–198.

[82] See, for example, *Escape from Woomera*, a game constructed by the Australian Escape from Woomera Team, a half-life game modification that puts the player into the role of a refugee trying to escape from a detention camp.

In Chapter 1, I presented Stahl's description of the viewing position interpolated by Google Earth, as having some of the attributes of a 'first-person shooter computer game'.[83] These attributes include the sense of surveillance, of being 'the eye behind a camera' which cruises above the earth and can zoom down and across to view more closely places chosen for interest and curiosity. When using this platform, there is certainly a sense of computer game play, as noted by Stahl. This particular 'game'-styled playing of Google Earth, however, is not driven only by the position of the 'first-person shooter'; it is also a world-building game.[84] Using Google Earth, we are able to build a world from our own searches for information – a beautiful world consisting of actual images of the Earth, which is itself annotated with spatial and time coordinates, and animated for different viewing styles if we wish. Nevertheless, this description of likeness between the experience of Google Earth and computer game play does not *per se* equate the platform with a gaming platform. It does, however, offer a way into understanding *how* personalized, so-called virtual worlds are when offered to and constructed by the user in collaboration with Google Earth. As discussed in Chapter 2, we *interact* with Google Earth in ways already available to us through and to the extent of our digital literacy. Stahl's detailed critique of vision via the technologies developed by Keyhole and further developed in Google Earth shows how these very attributes of 'flyby' and 'first-person shooter game' aesthetics also contribute to constructive and communal ways in which the application offers opportunities for interacting with information via Google Earth technology. Such attributes are certainly also associated with the immersive nature of the environments that we create when we are searching for and thereby constructing a sense of place and time during individual private uses of Google Earth.

One of the problems in representing the subjective experience of play that emerges from immersion in such an interactive site as Google Earth can be addressed, I suggest, by the process of 'thick description' of individual experiences. An example of this process is Lisa Parks' (2013) description of her own on-the-ground fieldwork for her research into Google Earth's

---

[83] Stahl, 'Becoming Bombs', 84–85.
[84] Currently popular world-building computer games still include *Everquest*, *The Sims* and *Spore*.

infrastructure. Parks applied mapping and photographic information gained from the Google Earth/FCCInfo Interface Layer[85] to find and visit the location of a transmission site in Santa Barbara, United States. Her account includes the following extract: 'I could hear the buzz of electrical equipment, feel the burning heat of the mid-afternoon sun baking the pavement, smell the scent of warm chaparral bush mixed with ocean air....'[86] As another example of using an affective personal account of matching the virtual with the actual via Google Earth, I am now going to present one of my own Google Earth journeys: 'Driving to the coast from Canberra via Braidwood (NSW Australia)'. I describe here my own experience in first person, in both reflective and diary modes.

'Flying through space above the earth, around the world': A fieldwork experience

The Kings Highway, NSW, Australia. This is the road between where I live in Canberra and the house I also live in as often as I can 'down the coast'. I find it difficult and tedious to drive. It goes down an escarpment in steep twists and tight bends, resolving into an even more frustrating looping pathway through the foothills, over a large river and on to the forests and beaches of southern New South Wales, Australia. After years of driving this road I wanted to see why it is like it is and thought I could do this by seeing it from above, as a pattern on the Earth. Post-2005, a friend referred me to Google Earth. So I began a personal exploration of a place on Earth that was already very familiar to me. Using Google Earth's 'directions' compass, I could run my mouse over this road at a low enough viewing height to see the forest on either side of it and the valleys and hills around which the road travelled (see Figure 3.1). I could also create a short, embedded (KMZ) video clip that very slowly 'drove' the road for me, keeping the point of view in the centre of the screen and swinging the road around and

---

[85] FCC = Federal Communication Commission, United States. For Cavell Mertz and Assoc. Inc.'s portal into this particular Layer, see http://www.fccinfo.com/fccinfo_google_earth.php (accessed 7 January 2014).

[86] Parks, 'Earth Observation and Signal Territories', 301.

back again through the road's bends. This movement in the screen to some extent mimicked the movement of my body as I steered my driving along the actual road itself.

This particular road is full of hair-pin turns, down a mountain and on to the sea at Batemans Bay NSW. I associate the landscapes of both the car trip and the coast with leisure and pleasure. But I do not like this trip itself if I am the driver. I find my negotiations of other traffic and the contours of road make it a difficult drive. I want to see why I find the road difficult to drive – to see its shape and therefore somehow to own it, to understand it differently and not just have to put up with my own very limited imagination. So I am using Google Earth to look at how this wretched road looks from the sky.

Suddenly, I have a new driving skill: 'driving the mouse in Google Earth', not quite so tricky as driving a car but now it feels like I am more in control – sort of – I tend to veer on and off the road into forested wilderness, but at least I can understand through my own mouse- driving mistakes some of the engineering problems in constructing this road through this particular terrain. Now I can also see the beginning and the end – the beginning and the destination of my journey.

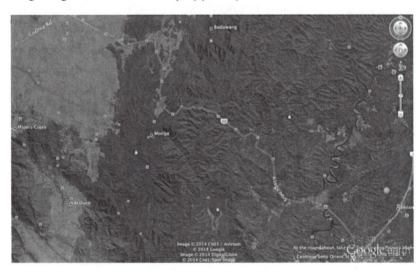

Figure 3.1 Screenshot of Hume Highway, Braidwood to Batemans Bay Road on Google Earth ©, http://www.google.com/Earth/ captured 5.31 p.m. EST, 25 August 2014. Google and the Google logo are registered trademarks of Google Inc., used with permission.

In his essay 'Tabula Rasa', Paul Virilio quotes Walter Benjamin on what Benjamin calls the 'force' of the road:

> The force of a country road differs depending on whether you are doing it on foot or flying over it in a plane. Only when you are travelling along the road can you learn something about its force.[87]

Now with Google Earth, I can experience another kind of force of this particular road; I can expand my experience of Benjaminian force through manually tracing the road over terrain I could not see from my car.

How to analyse this experience? Here I can draw on Benjamin again, whose words well and truly predate Google Earth but who spoke of changes in perception arising from new technologies. For example, Benjamin's idea of 'shock' evoked by the close-up images of cinema and the chaos of industrialized cities has direct relevance to my driving experience, in the sense that Google Earth allows me the shock of seeing a familiar road from above while simultaneously making possible a utopian but sensual perception of times and places in the world that are hitherto unfamiliar. My own personally constructed journey of the Braidwood coast road led to another journey. This time I lost and found my way across the Earth to Africa – recalling another quote from Benjamin in Virilio's essay:

> Not to find one's way around a city does not mean much. But to lose one's way in a city, as one loses one's way in a forest, requires some schooling.[88]

My 'driving' journey was also my first-time use of Google Earth. I followed up this journey with a look around the rest of the world – and I found the Global Awareness Layer 'Crisis in Darfur', marked with icons of flame, cameras, words and film-making. This layer is the case study which I examine in closer detail in the next chapter.

---

[87] Walter Benjamin, 'Berlin Childhood around 1900', in *Walter Benjamin: Selected Writings*, vol. 3, 1935–1938 (Cambridge, MA: The Belknap Press of Harvard University Press, 2002), 352, in Paul Virilio, 'Tabula Rasa', in *City of Panic*, trans. Julie Rose (Oxford, NY: Berg, 2007), 2.

[88] Benjamin in Virilio, 'Tabula Rasa', 1.

The cultural performance of Google Earth

Visualisation is seen to be something that happens in the mind of a human, not on the screen of a display.[89]

These words appear in the online description of the final report of the NATO IST-013/RTG-002 RTO-TR-030 – Visualisation of Massive Military Datasets: Human Factors, Applications, and Technologies. They draw attention again to the necessity of matching descriptions of people's abilities, intentions and digital literacy with how we describe their use of a platform like Google Earth. Ethical issues concerning the usage of Google Earth circle around the common property of most if not all current digital media: the knife-edge dilemma that exists in negotiating the divide between private and public domains of activity – corporate commodification versus individual enterprise – the interest of corporate and state-owned entities and the interests of the individual, smaller interest groups and NFP organizations. This dilemma has existed to some extent since word was printed on paper, but with the coming of digital media it has become much more easy to note a flow of power towards the individual person, even if this 'flow' is turbulent and in no way resolved nor even easily discernible. Crampton describes some of the social consequences and questions that arise because of 'distinctly public and citizen-oriented map-making efforts' involved in current digital mapping processes via remote sensing. He says that one important question involves 'the critical evaluation of the geoweb and whether it requires renewed map literacy or education'. He says that '[a]s with any technology, the particular systems of power and surveillance are unavoidable'.[90] Part of the answer then to the question 'why does everyman use Google Earth?' can draw on Baudrillard's claim (noted earlier in Chapter 2) that people who are disempowered by some process, either social or environmental, can nevertheless cling to illusionary signs of having power:

---

[89] Online Description of 'RTO-TR-030 – Visualisation of Massive Military Datasets: Human Factors, Applications, and Technologies', *Scientific Publications of NATO Research & Technology Organisation*, Tuesday 1 March 2001, http://nato-pubs.ekt.gr/NATORTO/handle/123456789/4541 (accessed 6 April 2014). ISBN 92-837-1066-5.

[90] Crampton, 'Cartography: Maps 2.0', 91.

And at the same time, another figure of power comes into play: that of a collective demand for *signs* of power – a holy union that is reconstructed around its disappearance.[91]

This idea plays into this question: are we being disempowered so much and in so many ways through digital technologies' offerings in surveillance, commodification and consumerism that we want to cling to any signs of power that we might claim as individuals? Or In order to be part of a new more egalitarian sharing of power, should we look past, whilst not ignoring, the dangers of using huge GIS Earth imaging programmes such as Google Earth: dangers that include loss of privacy, new kinds of colonialism and state control, new forms of corporate exploitation? Jason Farman's approach is to look closely at new forms of user agency. I think we can also find some answers by stepping back a little from structures of power, technologies and user experience. Then we can consider the whole spectrum of activities based on the phenomenon of Google Earth as contributing to a particular 'cultural performance'. Here I use this term not in Victor Turner's specific sense of 'social drama',[92] as a society's redressive performance towards resolving some kind of crisis (as specifically engaged with in Chapter 6). Nevertheless, it is useful to describe Google Earth as a cultural performance in order to draw attention to its attributes as a performative text. To quote Turner in part of his description of ritual performance:

> In the sense that the performance is often a critique, direct or veiled, of the social life it grows out of, an evaluation (with lively possibilities of rejection) of the way society handles history.[93]

In this book, I am describing Google Earth as a text whose existence includes performances of its making and reception. Parks (2013) frames her fieldwork in terms of performance, as 'creative mediation' – a concept that she drew

---

[91] Baudrillard, *Simulacra and Simulation*, 23.

[92] This concept was introduced by renowned anthropologist Victor Turner as part of his wider theory of 'social drama' – a concept of conflict, narrativity and process in social life that he began to develop through his early anthropological work on Ndembu ritual in Africa. As the third stage of 'social drama', cultural performance can include literature, theatre, film, both religious and secular rituals and sport. See Victor Turner, *The Anthropology of Performance* (New York, NY: PAJ Publications, 1986), 20–35.

[93] Turner, *The Anthropology of Performance*, 22.

from work by professors of new media and communication Sarah Kember and Joanna Zylinska. These authors ask this question:

> if we are saying that the events we have looked at are, to differing extents and in different ways, performed through their mediation, then how should we respond to them in our critiques?[94]

Similarly to Parks, I offer my own performance of Google Earth as creative mediation through which Kember and Zylinska claim it is possible to examine what they call the *'lifeness* of media': how critically experiencing media itself can bring new knowledge about 'unprecedented connections and unexpected events'.[95] This kind of critique involves performances that are connected by the very fact that they occur simultaneously in the virtual and the actual, the written and the actual, the representation and the phenomenon that is represented. Turner's idea, as does Kember's and Zylinska's, contributes to an understanding of performance as a mode of cultural critique. In this chapter, I looked at my own 'creative mediation', my road journey by both computer mouse and car. In the next two chapters I describe other cultural performances that are available through Google Earth and how associated performances of compassion might also be possible and valuable outcomes of journeying in Google Earth.

---

[94] Sarah Kember and Joanna Zylinska, *Life After New Media. Mediation as a Vital Process* (Cambridge, MA: MIT Press, 2012), xvii.

[95] Zylinska, *Life After New Media*, xvii.

4

'Crisis in Darfur': A Google Earth Outreach Case Study in Affect and Compassion

This chapter presents my primary case study for researching humanitarian intervention via Google Earth. My premise is that the affective and pragmatic conscious state of compassion – a particular kind of knowledge through compassion – is potentially at least, available via the very available interactive Google Earth website. As Parks notes, there is a significant range of possibilities for 'dual use' of Google Earth as a tool for affective understanding:

> [F]ew have considered how Google Earth builds upon and differs from earlier global media formats and how it structures geopolitics as a 'domain of affect', particularly when used as a technology of humanitarian intervention.[1]

I propose here that the reception and production of Global Awareness Layers such as 'Crisis in Darfur' offer a new kind of portal for experiencing compassionate understanding of other people, other places and other moments in time than the present. Before directly addressing 'Crisis in Darfur', I trace some concepts that are relevant to understanding this site: ideas of the public good, human rights activism via digital media and the affect of compassion itself. It is also necessary to acknowledge that the site did not appear from nowhere: there is a wide range of activism for Darfur and Sudan that occurred both before and after Crisis in Darfur was created.[2]

---

[1] Parks, 'Digging into Google Earth', 3.
[2] For an account of activism via acts of witnessing, both online and offline, see Leshu Torchin, *Creating the Witness. Documenting Genocide on Film, Video, and the Internet* (Minneapolis: University of Minnesota Press, 2012), Chapter 5, 'iWitnesses and Citizen Tube: Focus on Darfur', 172–220.

My general argument in this chapter extends outwards from Parks' comments in 2009 as quoted earlier. My closer argument develops around the following thought: while some uses of digital technologies clearly offer new ways to act in and interact with the world and new ways to perceive and experience the world, these new kinds of action and perception include a quiet mode of individual experience that happens at the human–computer interface (HCI) of our personal computer screens. This personal, highly subjective experience allows in turn a compassionate connection with those we encounter in cyberspace. Through analysing our experiences in Google Earth's Global Awareness Layers, we might then be able to describe, in Turkle's terms, another aspect of the 'new kind of sociality' that is emerging in the era of digital technology.

An affective process of immersion in cyberspace

In order to make a case for considering a personal computer as a viaduct or tool for experiencing the affective state of compassion, more needs to be said first about the ways in which space can be experienced via the personal computer. In previous chapters, I have noted current understandings of cyberspace as a convergence between physical space in the actual world and virtual space, particularly space that exists as a result of what is represented on the computer screen. The process of experiencing this convergence is called immersion. In digital studies, this term is used to denote what happens to us, how we feel where we are in space when this convergence occurs. The convergence itself and how it happens need also to be carefully described.

The process of algorithmic modelling that creates the virtual space offered by the personal computer is firmly linked currently to the concept of simulation. In Chapter 2, I discussed this term in the context of Baudrillard's idea of a hyperreal. Cubitt in his book *Simulation and Social Theory* (2001) calls his Chapter 3 'The Poetics of Pessimism'. In that chapter he traces various theories of simulation and their history. He notes the inherent pessimism in Guy Debord's philosophy of the spectacle, Virilio's concern with shifting relationship between time and representational space and the problems set out by Umberto Ecco in his theory of simulation and hyperreality. I suggest here, however, that the idea of simulation can be understood further, and more usefully perhaps when talking about Google Earth, to more generally mean a representational form that is very close to that of an event, object, landscape, human or other life form that exists as itself in the space of the actual world. No matter how 'real' a simulation seems, it is still a representation, although a particular kind of representation.

In this chapter, I draw on how new media theorists Jon Dovey and Helen W. Kennedy interrogate the role of simulation in computer game play. They use a useful long quotation from Friedman (1999) to describe what kind of representation practice is involved in the process of simulation. Friedman's words are particularly apt for thinking about spatial simulations in Google Earth:

> Representing flux and change is exactly what a simulation can do, by replacing the statis of two or three dimensional spatial models with a map that changes over time to reflect change. And this change is not simply a one way communication of a series of still images but a continually interactive process. Computer simulations bring the tools of narrative to mapmaking, allowing the individual not simply to observe structures, but to become experientially immersed in their logic.[3]

If we keep hold of the idea that simulation is nevertheless a style of representational practice, then it is possible to name the differences between simulation and the actual, in terms of representation. Masa Hiro Mori's 1970 thesis of an aesthetic he calls 'uncanny valley' offers an explanation why simulations, animations or robots that closely resemble the human form, can cause great discomfort or even revulsion in the observer.[4] The operative word here is 'resemble'. The simulation is *not* the actual, no matter how closely it comes to disguising itself as a phenomenon that exists in the actual

---

[3] Ted Friedman, 'Civilization and Its Discontents: Simulation, Subjectivity, and Space', in *On a Silver Platter: CD-ROMs and the Promises of a New Technology*, ed. Smith Greg (New York: New York University Press, 1999). As quoted in Jon Dovey and Helen W. Kennedy, *Game Cultures: Computer Games as New Media* (Maidenhead: Open University Press, 2006), 11.

[4] Wikipedia's entry 'The Uncanny Valley' gives a clear account of the concept, its history and application in animation: http://en.wikipedia.org/wiki/Uncanny_valley (accessed 29 January 2014).

world. It does of course exist in the actual through its primary function as a representational phenomenon that offers us information about the actual world. Simulations also exist within the world of cyberspace. It is how they operate as representations that are uncannily close to how we experience the actual world that is interesting here; and it is useful to remember at this point that Google Earth is constructed as a representation, as an animation that models the actual. The differences and how they are glossed over between representations and what they represent (signifier and signified in semiotic terms) can fool us into entering into and immersing ourselves in the 'as if' world of a representation – a world where we consider the actual world 'as if' it exists just as it is represented to us.

In the context of Google Earth, we enter into a representational world that is created through a process of algorithmic modelling of the actual: digital simulation. But I suggest that we are not (only) fools when we enter these worlds, these make-believe or 'as if' worlds. These digitally simulated worlds are also venues for speculation, for exploring the world through 'as if' 'what if' simulated scenarios that model the actual for us. The worlds shown to us through representational practice are subjunctive, 'as if' worlds. To understand information they might contain, we need to match up their content with what we know about our own experience of the present and the past, and even with what we think might exist in other people's present. This 'matching up', of our own and others' experiences with virtual worlds in cyberspace, can then be used to estimate possibilities and probabilities for the future.

Farman (2012) writes on how ideas of the virtual and of simulation, together with the distinctions (or lack of them) that we make between them, now 'fail us in important ways'. He says that they have come to stand in for each other. He suggests that the source of this failure can be followed back to Plato's worries about representational artistic practices obscuring the 'real', and to some extent, his discussion recalls Baudrillard's conceptualization of a dystopian hyperreal. Farman, however, moves beyond the idea of representation as *replacing* the 'real', by describing the experience of the virtual as 'sensory inscribed',[5] as 'an experience of multiplicity ... of layering,

---

[5] Wikipedia, The Uncanny Valley, 36.

and the constant interplay that bonds the virtual and the actual together is the pleasure of virtuality'.[6] This pleasurable (or not) experience of virtuality is not of course confined to the digital HCI. Both now and in the past, virtuality has been associated with the term 'immersion': to say that someone is immersed in a book or a film or a story or music, is to say that the person's mind and bodily attention is so focused on the world they are immersed in that they are no longer 'in touch' with the 'outside world'. Their 'inside world' is built from messages and representations, sounds and images from the virtual, and these all marry with personal and cultural memory, history and reflection to create a very convincing, if briefly, internalized world.

So although our immersion into virtual worlds has a long and continuing history in literature, oral storytelling, film and visual art, we are using digital technologies to create a new kind of virtuality. It is new because of its potential for explicitly interactive engagement that is combined with vivid visual and audiovisual representational practices – the latter being so vivid that they can sometimes create anxiety about representation replacing the real. And if the word 'actual' replaces the word 'real', it becomes easier to understand that virtuality exists, of itself, in the actual world.

There is still, however, an undercurrent of thinking about the virtual as though it exists in opposition to the 'real': the virtual does not belong to the real world – it is 'unreal'. But if we embrace the virtual as happening in the actual world, it makes sense to understand virtual worlds as wonderful places for exploring the actual world by using the thought domain of 'as if'. We can explore within this domain the fictional and non-fictional representational practices in many kinds of formats. The virtual then is a place that is part of the actual world.

Before going on to talk further about the role of interaction in how we immerse ourselves in digitally constructed virtuality, I need to describe the meanings I am giving to the words 'place' and 'space'. In general, people talk about virtual 'space' when they are describing digital virtuality. The use of the term 'space' is indeed a good term for inferring, connoting the kinds of loneliness and alienation that we might experience when we

---

[6]   Farman, *Mobile Interface Theory*, 38.

imagine ourselves floating around in an indifferent world, unsympathetic and unknowing of our individual needs and ambitions. 'Space' is not 'homey'; it is not even recognizable to us as something we can think about except as a set of dimensions described to us by experts (scientists).

'Place' is space that we *can* call home. We individually identify or associate ourselves with a specific space when we describe it as a place. This is especially interesting when we think about digitally produced social media such as Facebook. Social media promote the digital space they produce – their own brand of virtuality – as a place where we can safely present ourselves to others as our individual selves. They sell us a 'place' in virtual 'space'. This distinction between place and space is important because Google Earth can also be said to offer us planet Earth as a 'homey' place, even as it allows us also to see its trajectory as it spins through the complex world of 'outer space'. This allows for a feeling of individual agency; we feel we can to some extent control our own place within the daunting spaces of Earth and the largely unknowable spaces beyond Earth. This sense of agency is not so much an illusion as a way into manipulating the world to some extent. Computer games rely on this desire for control and manipulate us for our own entertainment, especially through 'first-person shooter' games.

Immersion via interactivity

As discussed in Chapter 3, the interactive experience of 'playing' Google Earth recalls to some extent the sensory pleasure of taking the 'first-person shooter' position in computer gaming. We can immerse ourselves in a book or a film; this immersion is not the result of passive reception. As noted earlier, our immersion in these formats, whether they are fiction or non-fiction, relies on our personal history, our memories and on specific bodily experiences that are relevant to the particular story embedded in the texts contained within a book, film or other format. Potentially, however, interaction with interactive digital texts can happen beyond our internalized personal worlds. Interactions between digital texts and the individual human user at the HCI are different from previous formats. So too is our immersion into the new kinds of narrative space that are offered by computers and interactive software programs.

In describing the kind of immersion that is involved in computer gaming, Dovey and Kennedy write that to experience it, 'We have to be doing'.[7] They say that while the commonly understood meaning of what it is to be immersed involves 'a loss of sense of self', and that in the interactive realm of computer gaming 'immersion in a game world is of a different order' – it is the outcome of 'intimate mental, emotional and physical engagement by the player with the game and the game technology'.[8] They claim that this kind of immersion is less about being consumed into a virtual reality and more about a state of intense concentration and focus on the activities required to experience the game,[9] including learning how to play the game, how to manoeuvre within the game space, how to use the controls, how to engage with other players and interestingly, puzzle solving. While Google Earth is not a computer game, the same kinds of activities are needed for obtaining information from the site. I suggest that one of the game-like 'puzzles' to be solved in Google Earth is how to use its search engine effectively and then how to make sense of the images that emerge from our searching. As with computer gaming, immersion in Google Earth depends on our levels of embodied, active interaction and engagement with its software.

Embodied immersion

The form of immersion that is evoked through interactivity at the HCI might also be described in relation to our experiences in the performance space of live theatre. The extreme responses that Artaud's extreme 'theatre of cruelty'[10] was designed to elicit provide a useful way into understanding how we flinch, turn away, laugh and cry for the horror, pathos and comedy

---

[7] Dovey and Kennedy, *Game Cultures*, 8.
[8] Dovey and Kennedy, *Game Cultures*, 8.
[9] Dovey and Kennedy, *Game Cultures*, 8.
[10] Whether or not Antoin Artaud's 'theatre of cruelty' is/was successful in theatrical terms is still being debated, for example, see Robert Vork, 'Things That No One Can Say: The Unspeakable Act in Artaud's *Les Cenci*', *Modern Drama* 56, no. 3 (2013): 306–326.

embedded in stories shown to us in theatre. Live theatre, with its explicit mimetic engagement between the bodies of performers and audience, is in fact a highly relevant trope for considering our reception of Google Earth, as I will be exploring in more detail in Chapter 5. To a greater or lesser extent, our mimetic responses to live theatre can affect our own bodies 'as if' the bodies of the actors are our bodies. Our immersion and its responsive consequences can be described in a metaphorical sense: we are 'touched' by their bodies and the world created via their theatrical performances. As we approach the matter of how compassion is available to us through our affective response to interactions with content shown on the screen of a personal computer, it is useful to explore also the ways in which we can understand the term 'touch' in the explicitly literal sense of consciously experienced bodily sensation.

Farman's discussion of virtuality and simulation that I noted earlier and also in previous chapters forms the basis for his theory of embodiment that he calls 'the sensory-inscribed' body.[11] His concept in turn is drawn from a theory of affect that draws the sense of touch – haptic experience – into the arena of textual reception. He draws on Merleau-Ponty's description of the act of interpretation as one that is related directly to our perception of our own bodies and he names the sense of touch as one that is necessary for constructing our overall sense of body space and movement. Whilst this description is focused on how the body constructs its sense of body space, Merleau-Ponty does say the following about the processes necessary for acts of self-interpretation. He notes how visual data can be taken in as a 'bodily event' – how we reveal ourselves to ourselves in the ways we interpret information that comes to us from outside our bodies:

> Here the 'visual data' make their appearance only through the sense of touch, tactile data through sight, each localized movement against a background of some inclusive position, each bodily event, whatever the 'analyser' which reveals it, against a background of significance in which its remotest repercussions are at least foreshadowed and the possibility of an intersensory parity immediately furnished. What unites 'tactile

---

[11] Farman, *Mobile Interface Theory*, 31.

sensations' in the hand and links them to visual perceptions of the same hand, and to perceptions of other bodily areas, is a certain style informing my manual gestures ...[12]

This linking of the sense of touch to the experience of interpreting the actual is an important one for understanding the idea that we can gain knowledge in the form of a reasoned response to embodied affective emotional states. Put more simply, we are considering here the possibilities for embodied knowledge via the largest organ of the human body: skin. We experience touch primarily through the skin as friction and movement through space. As noted earlier, touch is a powerful tool, both metaphorically and literally, for describing how we use the virtuality of representational spaces to gain knowledge about the actual world.

New media philosopher Mark B.N. Hansen draws on Brian Massumi's work that in turn draws on Deleuze's film theory regarding the time image and the movement image, to investigate what Hansen calls 'Affecting Haptic Space'.[13] He uses Massumi's idea of proprioception to describe 'what happens in the body when it acts on itself in order to produce an internal haptic space'.[14] He quotes Massumi's definition of proprioception as follows:

> Proprioception folds tactility into the body, enveloping the skin's contact with the external world in a dimension of medium depth: between exodermis and viscera ... the faculty of proprioception operates as a corporeal transformer of tactility into quasi-tactility ...[15]

Hansen goes on to describe how this internal haptic space combines with sight and the awareness of movement via sight, to produce an overall perception of something:

> Movement-vision and proprioception, in short, comprise the visual and corporeal dimensions, respectively, *of a single perceptual experience*.[16]

---

[12] Merleau-Ponty, *Phenomenology of Perception*, 174.
[13] Hansen, *New Philosophy for New Media*, 227.
[14] Hansen, *New Philosophy for New Media*, 227.
[15] Brian Massumi, 'The Bleed: Where Body Meets Image', in *Rethinking Borders*, ed. J.C. Welchman (London: Macmillan Press, 1996), 30–31. As quoted in Hansen, *New Philosophy for New Media*, 228.
[16] Hansen, *New Philosophy for New Media*, 228.

Hansen interrogates digital art as a specific way to explore 'the bodily dimension of image perception' as it occurs in digital media.[17] For example, he examines viewer participation/interaction that is required for viewing Alba d'Urbano's *Touch Me* (1995).[18] In this work, a viewer sees the face of a woman and is invited to touch it (i.e. the monitor screen that shows the image, the HCI). The viewer's touch results in the image of the face disintegrating, and at the end of this destruction the screen displays a live video of the user during this interaction with the work of art. This work of art explicitly demands a haptic connection between the computer screen and a human body. Hansen inquires closely into Massumi's definition of affect as '[t]he perception of one's own vitality, one's sense of aliveness, of changeability'[19]; affect as

> an ability to affect and be affected. It is a prepersonal intensity corresponding to the passage from one experiential state of the body to another and implying an augmentation or diminution in that body's capacity to act.[20]

Both Hansen and Massumi's philosophies of affect are complex, and it is well beyond the scope of this book to engage with them in depth. However, it is useful to briefly address Massumi's distinction between affective vitality – the 'ability to affect or be affected' in an event – and emotions as categories of bodily responses to spaces of content during the event. He clearly distinguishes between affect and emotion:

> An emotion is a subjective content, the socio-linguistic fixing of the quality of an experience which is from that point onward defined as personal. Emotion is qualified intensity, the conventional, consensual point of insertion of intensity…into narrativizable action-reaction circuits, into function and meaning. It is intensity owned and recognized.[21]

---

[17] Hansen, *New Philosophy for New Media*, 141.

[18] Hansen, *New Philosophy for New Media*, 141.

[19] Brian Massumi, 'The Autonomy of Affect' (1995). http://www.brianmassumi.com/textes/ Autonomy%20of%20Affect.PDF (accessed 30 January 2014) as quoted in Hansen, *New Philosophy for New Media*, 227.

[20] Brian Massumi, 'Notes on the Translation and Acknowledgments. Pleasures of Philosophy', in *A Thousand Plateaus. Capitalism and Schizophrenia*, eds Gilles Deleuze and Félix Guattari (Minneapolis: University of Minnesota Press, 1987), xvi.

[21] Massumi, 'The Autonomy of Affect'.

Massumi's definitions and descriptions are useful for considering Google Earth because they can be used to locate this platform not only as a text with an aesthetic to be interpreted but also as a performative event that happens as a phenomenon in itself and which includes those of us who interact with it in any way. This is not to say 'we are Google Earth' or that we live in a 'googlized' world, but it is to say that we are engaged in being in the event of Google Earth with all its complexities and consequences when we engage with Google Earth. Yes, we can communicate to ourselves and to others about this event via the emotions that emerge from our participation in the event and our responses to these emotions. In other words, we need to interpret our own situation when using Google Earth not only by considering how the platform is affected by us or by how it affects us. Rather, we are part of the event itself. We are situated within the event of Google Earth. Our 'interaction' with its aesthetic, its textual content is 'action' within the event of Google Earth.[22] We are part of its performance, as bodily individuals, yes; but also as players, performers of a vital phenomenon that constantly exceeds its assumed potential for enabling communication and human experience. So we can say now that to talk about Google Earth as an interface between the virtual and the actual is to ignore the complicated implications of its overall affect that we experience through immersion into the virtuality of its animations and simulations.

In Massumi's terms, the sensually and often pleasurable experience of bodily space that we experience on our immersion in Google Earth is part of the affect of Google Earth. We then can experience culturally described, categorical emotions and describe how they make us feel. So drawing on this acknowledgement that Google Earth is an event that can affect us, and by immersing ourselves in its content, I propose that we can feel personally the emotion of compassion as a response to content shown in Google Earth. My following discussion of compassion, however, does not leave compassion to be understood as only an emotion and personal feeling.

---

[22] My understanding of these ideas is drawn from my reading of Brian Massumi, especially in his writing on the 'Aesthetico/Political' in the context of performance in dance and music, in *Semblance and Event. Activist Philosophy and the Occurrent Arts* (Cambridge, MA: MIT Press, 2011), 152–156.

Compassion

Compassion is both a complex concept and praxis: a state of mind. It includes primarily a willingness to be open to others and to acknowledge others as being co-creatures; emotions – sometimes anger, fear both for ourselves and, drawing on Levinas, fear for the other and of our responsibility for the other[23]; often empathy – another emotional feeling that derives from a strong, sometimes explicitly embodied identification with the experience of others; and the actions required to access our potential for compassion – various and always sensory. Compassion, in terms of affect, is a culturally defined emotion. When we say we 'have compassion for …' we mean that we are ourselves feeling the emotion our culture calls compassion.

There are also hazards, however, in thinking about compassion as producing a simple kind of personally experienced feeling. As understood in the popular sense, compassion can have negative connotations that extend onwards to places and beings that we might feel compassion for. For example, the images of a whole and navigable planet Earth such as those offered by Google Earth reinforce this sense of the wholeness of our world. The ideology that draws on the Gaia principle[24] provides a good example of how we humans can imagine ourselves to 'feel' for our planet as a whole biosphere that includes the animate and the inanimate; and how simultaneously it is to a large extent beyond our control; our influence over how Gaia continues to change. I suggest that this now popularized notion of Earth as Gaia[25] underlies some people's lack of understanding of the science of climate change. This science is sometimes disparaged via an overall disparagement of emotionally based 'feelings' that are thought to indicate a state of intellectual delusion, irrationality and overall weakness of will. 'Feeling sorry for someone doesn't help anything' is a common saying that conflates the idea of feeling sorry with compassion and reduces it to a useless set of actions and reactions.

---

[23] Emmanuel Levinas, 'Ethics as First Philosophy', in *The Levinas Reader*, ed. Seán Hand (Oxford: Basil Blackwell, 1989), 82.

[24] See 'Gaia Hypothesis', *Wikipedia*, http://en.wikipedia.org/wiki/Gaia_hypothesis (accessed 11 February 2014).

[25] See, for example, the anime film *Final Fantasy: The Spirits Within* (Hironobu Sakaguchi and Motonori Sakakibara, 2001).

I want now to introduce more forcefully the idea that compassion, what is often mundanely called the act of 'feeling sorry', is an act valuable of itself. This act is not simple and I propose to use the term 'compassion' to challenge the banal 'common sense' point of view that all the many and various layers of the 'feeling sorry' can be condensed into a non-productive, self-deluding gesture of a particular (in the West) Christian moral position. The word to hold on to here is the term 'act', and it is important to use this term literally. Acts of compassion, of feeling sorry for or towards someone or something that we can see in live space or in the mediated space of a computer screen, are haptic, acts of touching with our eyes, imaginations and memories and literally with our hands on the computer mouse. These acts have consequences – they are not irrelevant to how the world as a whole works politically, socially, as the whole environment for how we exist as humans.

The emotion that we call compassion always has an outcome because it is also a state of recognition, of knowledge. The embodied experience of compassion that I have just described educates us about the world and so has the potential to make us more politically aware and to provoke us to take some kind of effective action in the realm of social justice. This quote from Levinas well articulates the relationship between what we think and what we know as embodied action:

> But in knowledge there also appears the notion of an intellectual activity or of a reasoning will – a way of doing something which consists precisely of thinking through knowing, of seizing something and making it one's own, of reducing to presence and representing the difference of being, an activity which *appropriates* and *grasps* the otherness of the known.[26]

Levinas reasons that 'Knowledge as perception concept, comprehension, refers back to an act of grasping'[27] and that this metaphor of grasping and seizure is a metaphor [that] should be taken literally. He goes on to describe how the 'immanence of the known to the act of knowing is

---

[26] Levinas, 'Ethics as First Philosophy', 76.
[27] Levinas, 'Ethics as First Philosophy', 76.

already the embodiment of seizure'.[28] Although a strong emotional feeling of compassion is not required for an outcome of effective political action, the consciously willed state of open-mindedness towards experiencing the emotion *is* required.

We do not have to be in the grips of a strong emotion 'to feel compassion', 'to have compassion'. It is this last phrase, 'to have compassion', that more accurately denotes the choice involved with entering into the realm of affect, the event that is Google Earth. When we have compassion for people and other beings that we see in Google Earth, we have already decided to open our minds to feelings of compassion that reach beyond that virtual space of the intellect. In Levinas' terms, we have grasped, taken for ourselves a state of mind that is open to embodied knowledge about something that is knowable through empathy, an embodied knowledge of those which we perceive as different to ourselves.

There is certainly an asymmetry in power and experience between us and people we can see and who cannot see us, but these people, as well as landscapes we know only through mediated images, can certainly touch us in the various ways described earlier. Images can engage us and therefore also have an effect on the ways we ourselves live our lives. For example, when we open the shades of a plane's window when flying over Afghanistan at 30,000 feet, we see the actual brown folding hills of this country and know at the same time that terrible conflicts of war are happening, hidden and yet marked by the topography itself. The kind of actuality that confronts us at these casual moments of viewing the world beneath us is both easy to apprehend and yet terrifying in the enormity of grief and sadness which might be a part of our perception of that topography and the human lives it signifies to us personally. Google Earth, on the other hand, is a visualization, a simulation of information gained via the touch of remote sensing satellites. Nevertheless, it can show the same kind of 'distant' image that we see beneath that impossibly flying plane, and it can include something more. We can humanize the topographies we see by revealing that which we can only speculate on and imagine as plane passengers flying over places of conflict. We can create layers within Google Earth.

---

[28] Levinas, 'Ethics as First Philosophy', 76.

Both experiences of 'flight' (simulated and actual) shake the boundaries of our physical selves, of where and how we exist in time and space. I suggest that these experiences of flight can unsettle us to the extent that we experience the other person and other place as a particular kind of encounter that is marked by feelings of the emotion compassion. Martha Nussbaum draws on Aristotelian philosophy to define the Western culturally described categorical emotion of compassion as

> a central bridge between the individual and the community: it is conceived of as our species' way of hooking the interests of the others to our own personal goods.[29]

In this sense, it is the emotion that names that feeling of temporary identification with someone else who is suffering what they do not deserve. In describing compassion as a state defined well via the writer and philosopher Rousseau's fictional character Emile, she quotes Rousseau (as Emile) as follows: 'To see it without feeling it is not to know it.' She immediately goes on to explain that at this stage of Emile's narrative trajectory 'the suffering of others has not become part of Emile's cognitive repertory in such a way that it will influence his conduct, provide him with motives and expectations...'.[30]

Nussbaum understands compassion as 'entering into...lives with empathy and seeing the human meaning of the issues at stake in them.'[31] Do we need to empathize so strongly then with people whom we see suffering that we feel pain ourselves? This is surely an ability few could have and probably none would want: is this ability necessary to experience compassion? Nussbaum answers this question by using the fiction of Rousseau's Emile once again: 'No such particular bodily feeling is necessary...we look for the evidence of a sort of certain thought and imagination, in what he says, and in what he does.'[32] Indeed, Nussbaum understands the emotion of compassion as necessary for developing knowledge of other people, both as individuals and as communities. She speaks of imagination. I suggest here that haptic,

[29] Nussbaum, 'Compassion', 28.
[30] Nussbaum, 'Compassion', 38.
[31] Nussbaum, 'Compassion', 53.
[32] Nussbaum, 'Compassion', 53.

interactive ways of perceiving the images we find in 'Crisis in Darfur' can provide a key into that faculty of imagination that resolves into the knowing emotion of compassion. Nussbaum describes compassion as a kind of knowledge accessed through feeling, not relying on the strength of that feeling but on its quality of being open to other people and places.

'Crisis in Darfur'

Before going on with this part of my discussion, I need first to point out that the 'Crisis in Darfur' layer has changed from the time of my original research to now. I look more closely at this process of change in Chapter 5 together with the historical context to the conflict that the site depicts. For now the reader needs to know that many links have now been deleted or redirected and new links have been made. My account of 'Crisis in Darfur' here relates to the site as it was created and maintained during 2007–2009 by Google Earth and the United States Holocaust Memorial Museum (USHMM). In the next section, I address my responses to and knowledge of the site through my primary research that took place in 2010. Whilst these responses remain valid, my overall research into the site sheds light on the tenuous state of authority of information that is embedded into the malleable format of a Keyhole Markup Language (KML) layer.

After I had investigated my own personal journey along a south coast road in New South Wales, Australia, I swung over the seas towards Africa, as part of my Google Earth trip around the world. I had clicked on the Google Earth Awareness Layers without knowing what they were. In the Sudan area in the north of Africa the bright flames of the site 'Crisis in Darfur' drew my eyes and, as I zoomed down, I began to unravel the confusion of images and stories contained within that site. My journey through the information of 'Crisis in Darfur' was even more confused than my first attempts to drive my mouse over the coast road southwest of Canberra. It is not an easy task to work through the various icons and find the images and words that are buried beneath them. My experience of trying to understand what I was seeing in 'Crisis in Darfur' marked the beginning of my work on Google Earth.

I have returned many times to Sudan through Google Earth and every trip is different. The numbers of displaced people overwhelm with their magnitude, and the photographs of the refugee camps ground this information fairly well. The images though always affect me strongly; the level of photography is very good and I find myself virtually face-to-face with people who have suffered atrocities and who are presented to me within that content of suffering. On the other hand, the documentary video clips that can be accessed via links to the USHMM site pull me back to myself as I listen to other people translating the suffering in voice-over, telling me for example, about the drawings children made in their refugee camps – drawings depicting their parents and families being massacred.[33]

The combination of written, translated testimonies of the victims themselves, together with illustrating imagery, always affects me the most, pulling me back into a dialogic space where they and I exist together – they, through their representation, their images and stories, and me, through my intellectual and affective knowledge of what has happened to them, of what they are describing to me as I sit in front of my computer screen. During the time of my finding and witnessing their trouble, they are in my live embodied space, my domestic space, and I have to deal with that. Later, I confront the question of how the knowledge that I gain from such human rights media through my engagement with Google Earth differs from that accessed through other media formats. How might this knowledge be different from that gained directly from activist websites, like that of the USHMM?

One difference is this: I instigated a search for something I did not know was there and then followed through with a focused investigation. I looked at this site with the same body that could drive the coast road on and off screen. I played Google Earth with my own body in order to understand places and situations I knew of previously either at a great distance (with little affective response) or at too close a distance (with a great degree of affect). So I asked myself, how could I deal with an affective knowledge of people with whom I had no live 'face-to-face' engagement? How could I name such

---

[33] To view the clip, 'The Smallest Witness' follow this link that I accessed on 29 September 2010: http://www.ushmm.org/genocide/analysis/details.php?content=2005-06-03. I discuss and analyse this particular clip more closely in Chapter 5.

knowledge and how far could I go in defining this interaction as a new kind of face-to-face engagement mediated by the disappearance of distance, both actual and perceptual?

I found the site because it was covered with a mass of yellow and red icons of flame that sometimes hid other icons of cameras, words and film-making. 'Crisis in Darfur' is not an easy site to access: its icons overlap and relay a sense of confusion – of chaos even – and so create a gesture of searching that is indicative of the situation represented: one of war and displacement (Figure 4.1).

After exploring this site as best I could during that first-time experience, I thought about the conjunction of my 'virtual' personal, domestic journey in my own homeland and the journey I took across to Africa shortly afterwards. This question emerged as significant: did each of these journeys imply an inherent way of knowing another person? In other words, was I able to transfer the sense of immediacy in time and space that I felt in my journey to the south coast of New South Wales in Australia, to my confrontations with the agonies shown in 'Crisis in Darfur'? While this question is unanswerable in a literal sense, I suggest now that the question itself provides some kind of answer. The question both describes

Figure 4.1 Screenshot of 'BURNING OF UM ZEIFA. This is the beginning of the burning of the village of Um Zeifa after the Janjaweed looted and attacked.' From Google Earth ©, USHMM © captured 19 September 2010. Google and the Google logo are registered trademarks of Google Inc., used with permission.

and speculates on how we can respond to Google Earth in the context of feeling and emotion. States of embodiment involve our actual cognitive and physical manipulations of personal computer technologies and usually, as Turkle notes,[34] also involve our physical, spatial isolation from those we connect with via these technologies. To rephrase my earlier question in the context of spatial isolation: how and why can we nevertheless make connections with the people represented on our HCI despite this physical isolation?

Drawing on my earlier discussion of affect in Google Earth and the embodied knowledge available through the emotional feeling of compassion, I now want to investigate 'Crisis in Darfur' for how it might prompt us to feel compassion for the people we see within its images. The human rights activist site 'Crisis in Darfur' was launched in 2007, only two years after the launch of Google Earth itself and was introduced in collaboration with the USHMM. In this long quote, the Director of Communications of the USHMM Andrew Hollinger explains 'How They Did It' in an article posted on a Google Earth Outreach page:

> Development began in earnest when an international volunteer organisation, the BrightEarth Project, was formed to explore how a new generation of mapping tools, including Google Earth, could empower citizens around the world to better defend vulnerable populations. ... Museum staff and Bright Earth volunteers worked for more than a year ... and created the first KML draft layers in early 2006 ... But without high-resolution imagery, presenting the data in Google Earth was only a slight improvement over traditional maps ... and between autumn 2006 and spring 2007 the Google Earth team updated large swaths of Darfur with high-resolution imagery ... With the data alone the user could see the big picture of attacks in Darfur but had no understanding of the local impact on each village and settlement. When the two sets of data were combined, each became more powerful.[35]

---

[34] See Sherry Turkle, *Alone Together. Why We Expect More from Each Other and Less from Technology* (New York, NY: Basic Books, 2011).

[35] Andrew Hollinger, 'United States Holocaust Memorial Museum Crisis in Darfur' on the Google Earth Outreach page: http://www.google.com.au/earth/outreach/stories/darfur.html (accessed 12 February 2014).

This collaboration marked the first of many collaborations that Google Earth Outreach undertook with various non-government organizations and not-for-profit organizations, human rights organizations, environmental activist organizations and governments. The high-profile collaboration with the USHMM in producing the 'Crisis in Darfur' layer was and I think still is Google Earth Outreach's most powerful performance of archive and advocacy. In Rebecca Moore's words soon after its launch in 2007:

> The reaction to this project has been immediate: it has stimulated extensive worldwide media coverage, traffic to the USHMM website has quadrupled, and reporters and human rights organizations have used the information in these layers to ask more pointed questions.[36]

Interestingly, Moore notes that linking people to the USHMM's own website is a significant role of the 'Crisis in Darfur' layer.

'Crisis in Darfur' has been and still is a complex, multi-layered document of the massacres, vast displacement of people together with the destruction of villages occurring in Darfur as a result of the 1983 civil war in Sudan; this particular level of conflict is called the Third Darfur Rebellion.[37] During this war, more than 300,000 people were killed and 2,500,000 driven from their homes in the Darfur region of western Sudan.[38] The Google Earth layer shows high-resolution photographs of more than 1,600 destroyed and/or damaged villages and the remains of more than 100,000 homes, schools and other buildings destroyed by Sudanese government and militia forces.[39] Through this site we gain access to stories, photographs, statistics and videos that are laid over/embedded in a vast topography of human destruction. Some icons introduce us to higher resolution shots of the Earth, zooming across landscapes of burnt villages and refugee tent cities. In 'Crisis in

---

[36] Moore, 'Raising Global Awareness in Google Earth', 3.

[37] For more detailed information on the history of the Darfur rebellions and the conflicts between northern and southern Sudan, see Natsios (2012). For a close history of the current third Darfur rebellion in all its complexity, see Chapters 7, 8 and 10. See also the Human Rights Watch web page: 'Failing Darfur' http://www.hrw.org/sites/default/files/features/darfur/index.html (accessed 3 February 2013).

[38] Hollinger, 'United States Holocaust Memorial Museum Crisis in Darfur'. http://www.google.com.au/earth/outreach/stories/darfur.html (accessed 12 February 2014).

[39] Moore, 'Raising Global Awareness with Google Earth', 2. http://www.imagingnotes.com/go/article_free.php?mp_id=97&cat_id=18&Udo (accessed 22 January 2014).

Darfur' we see small iconic images of cameras (for photographs), quotation marks (for written testimony), clapperboards (for audiovisual material), small blue pyramids (for refugee tent cities) and different coloured flames to denote the various levels of devastation of specific villages. These various icons act as hyperlink points to other sites of information that occur either as part of the Google Earth site itself (links to photographs and text) or to other websites.

The site is very 'messy' at first glance. There is so much embedded information that the icons slip over and under each other and it is sometimes impossible to find the same link again during consecutive viewings. After examining 99 per cent of the site in 2010, it was clear that there are over 230 primary links denoted by as many icons. These links are associated with particular villages or refugee camps. There are also thirty testimonies from people who were victim to the *janjaweed*, the government-backed armed militias that ravaged the people and countryside of Sudan during that phase of the civil war.

The information found via these links is primarily from the USHMM website, including these internal links and updates: 'Mapping Initiatives: "Crisis in Darfur" (2009 Update)', 'Mapping Initiatives: Be a Witness', 'Speaker Series', 'Responding to Genocide To-Day', 'Take Action' and 'Who Is at Risk?' Other linked sites include those created by film-makers, Doctors Without Borders, the Universal Declaration of Human Rights, Physicians for Human Rights, the United Nations Environment Program, Global Grass Roots, United Nations High Commissioner for Refugees – the United Nations Refugee Agency, the United Nations Office for the Coordination of Humanitarian Affairs and Amnesty International.

Institutions and sometimes individuals are named as sources of the information embedded beneath the icons. The nature of this information is sometimes rather drily presented, through maps or the statement of statistics. More often, the information comes through images and witness testimonies of great suffering. This is an example of the latter:

The attack took place at 8am on 29 February 2004 when soldiers arrived by car, camels and horses. The Janjaweed were inside the houses and the soldiers outside. Some 15 women and girls who had not fled quickly enough were raped in different huts in the village. The Janjaweed broke the limbs (arms or legs) of some women and girls to prevent them from escaping. The Janjawid remained in the village for six or seven days. After the rapes, the Janjawid looted the houses.[40]

These testimonies, together with photographs and films, constitute what Rancière calls 'intolerable' images.[41]

The 'intolerable image'

The responses of 'pain and indignation' to the intolerable image are directed towards what is deemed abject to society, 'too intolerably real to be offered in the form of an image'. Rancière goes on to say that 'This is not a simple matter of respect for personal dignity.'[42] He writes how this denial of vision is not so much a matter of respect for the embodied human who is represented as disempowered through the violence of pain, but rather as a matter of damning such representation because this representational practice is embedded in a regime of vision which is the same as his example of the 'view of the dead child in the beautiful apartment'.[43] In this example, the photographically indexical image of a brutalized child becomes a work of art that in turn can become part of the décor of an expensively appointed domestic living room.

Rancière then describes how a specific authoritative voice of society emerges to tell us that it is immoral to view such images without taking action to reverse the wrongs that are represented within them: 'Action is presented as the only answer to the evil of the image and the guilt of the spectator.'[44] He shows how such moral action is made impossible, however, by the spectator's guilty immersion in the 'false existence' of the 'unrepresentable'. Rancière

---

[40] This testimony lay under the icon over the village of Um Baru. It was from a '30-year-old woman'. Information was provided by Amnesty International and the USHMM. The date of this particular page set-up was not given when accessed on 3 February 2011.

[41] See Chapter 1 for my first discussion of Rancière's concept of the intolerable image.

[42] See Chapter 1 for my first discussion of Rancière's concept of the intolerable image.

[43] Rancière, *The Emancipated Spectator*, 85.

[44] Rancière, *The Emancipated Spectator*, 87.

thereby distinguishes between 'the intolerable image' and what he names as a shift to the 'intolerability of images'.[45] So it becomes an issue yet again of what can be seen and what cannot be seen, although this time it is a 'lose–lose' situation for the would-be spectator. In the digital age, it is possible perhaps to draw an analogy between the 'expensively appointed apartment' and the comfortably appointed viewing space of the personal computer, and to deny the morality of looking at pictures of suffering because they come from the same envisioned regime of the spectacle.

When dealing with the subject matter of human rights abuse and war, however, the unrepresentable is always in play under the guise of the 'unthinkable' and that is the challenge offered by 'Crisis in Darfur'. As we search and find by happenstance some of the terrible things humans do to each other, we are confronted with having to think and imagine at a new level. To form a constructive political act of seeing and watching without fearing associated feelings of voyeuristic fascination, we need to use this emotional imagination to embrace the people and places that are *inside* the image.[46]

Framing the 'disappeared'

It is to the stranger that we are bound, the one, or the ones, we never knew and never chose. To kill the other is to deny my life, not just mine alone, but that sense of my life, which is, from the start, and invariably, social life.

Judith Butler[47]

With these words, Judith Butler cuts through any prevarication inherent to the morality, or what I would call a 'false' morality in this context, which denies our looking at the intolerable suffering of others, when such a denial disguises a desire 'not to think', not to contemplate the suffering of others. While Butler is talking about the literal killing of people in war, her emphasis is that if we

---

[45] Rancière, *The Emancipated Spectator*, 84.
[46] For a comprehensive discussion of how images can style spectators as political agents, see Lilie Chouliaraki, 'The Media as Moral Education: Mediation and Action', *Media Culture and Society* 30, no. 6 (2008): 831–852.
[47] Judith Butler, *Frames of War: When Is Life Grievable?* (London: Verso, 2010), xxvi.

kill another we also kill the 'sense of my life which, is from the start, and invariably, social life'.[48] Butler's words speak in turn and directly to a social need to affirm life and the links we have through this life to other people, via whatever mediation we have available.

We look at the stranger and draw them into our space as they draw us into theirs. This space then becomes a dialogical one in which, using Michael Holquist's words,

> all meaning is relative in the sense that it comes about only as a result of the relation between two bodies occupying *simultaneous but different* space [italics mine], where bodies may be thought of as ranging from the immediacy of our physical bodies, to political bodies and to bodies of ideas in general (ideologies).[49]

Holquist's understanding of Bakhtin's dialogical communication clearly refers to the possible creation of meaning between physical bodies 'occupying *simultaneous but different* space'. In this sense, it is 'far distance' which disappears. The people we see/hear through the textual practices of film and photography are present, not absent in this communication space. Their images refer to their presence, not absence, in the joined space through which we engage our social life – a life which includes them. How then to explain why distances of the 'far away' still inform so powerfully society's understanding of computer-mediated communication? A destructive relativist response still exists yet in our viewing of the 'stranger' who suffers, that is 'that is their business, not mine' closely followed by the fearful response, 'it couldn't/ shouldn't happen to me, I am different.' Interestingly, Nussbaum's definition of compassion includes such fear as a constructive affective force.[50] But what are the aesthetic reasons that might contribute to this fear as a destructive response to the 'other'?

In his essay 'Kriegstrasse', Paul Virilio writes about the 'telepresence of terror'[51] and proposes that terrorism in the twentieth century was marked by the trope of disappearance. While the last century was marked by an

---

[48] Butler, *Frames of War*, xxvi.

[49] Michael Holquist, *Dialogism: Bakhtin and His World* (London: Routledge, 1990), 22.

[50] Nussbaum, 'Compassion', 35–36.

[51] Paul Virilio, 'Kriegstrasse', in *City of Panic*, trans. Julie Rose (Oxford and New York: Berg, 2007), 55.

'*aesthetics* of disappearance', the twenty-first century, he argues, is now developing an '*ethics* of disappearance'.[52] Virilio suggests that today society is striving for immediacy, for speed and mass affect to which end individual morality and emotional affect will be lost – a pessimistic view indeed. But if we accept his thesis here, what is actually disappearing at such speed? Are we disappearing from each other as embodied beings?

Perceptions of distance between people are certainly changing. Online communication allows people far away to communicate as if in the same live space and time, echoing the immediacy of telephony, the accessibility of Web 2.0 and email. There are also huge increases of this kind of 'immediate' communications traffic, in which far is experienced as near. Has the idea of 'far' subsequently disappeared, then, in the realm of interactions between individual people? And has the distance of cultural difference disappeared? What has disappeared, I suggest, is the perception of distance between the far and the near, and that the difference between near and far is no longer of the same significance for how we now understand the world around us. I also suggest, however, that the 'distance of difference' is as strong as ever – contrary to Virilio's teleological account of this century's 'ethics of disappearance' and his prediction of a homogenized future.

Cultural difference may be disappearing to some extent in the globalized world, but I suggest that this is happening more slowly than we think it is. But it is far too early to say that the social and perceptual apparatus for individual experiences of subjectivity has disappeared. It is changing, not disappearing. In McLuhan's words,

> Our sensory modes are constituents, not classifications. I am simply identifying modes of experience. We need new perceptions to cope. Our technologies are generations ahead of our thinking.[53]

McLuhan explains that by investigating how technologies affect our understanding of ourselves, we are in fact 'dealing with the present as the future'.[54] And this is a key point in considering new media texts: digital media

---

[52] Virilio, 'Kriegstrasse', 56.
[53] Marshall McLuhan, 'A Dialogue: Gerald E. Stearn and Marshall McLuhan', in *McLuhan Hot and Cool*, ed. Gerald E. Stearn (Harmondworth: Penguin Books, 1968), 336.
[54] McLuhan, 'A Dialogue', 337.

algorithmically models the actual world as we imagine and speculate it to be; speculation is the very material of digital technologies. Such speculation is mediated *human* activity – we still program the computers that program the computers that ... etc. The 'new perceptions' that we need to cope with both the present and the speculative future require a fundamental change in our understanding of who we are as humans. Our knowledge of ourselves grows out of these changes, perhaps towards an acknowledgement that 'out of sight' can no longer be 'out of mind', that our ideas about who we are no longer constrained by what we know, but by what we *can* know.

Face-to-face

Kathryne Hayles expresses the following *cri de coeur* for humanity's existence in the age of the *posthuman*, as an existence that happens both within and beyond those networked relationships instigated by computer technologies:

> Other voices insist that the body cannot be left behind, that the specificities of embodiment matter ... bodies can never be made of information alone, no matter which side of the computer screen they are on.[55]

With these words she calls for an understanding of a concept called the *posthuman*, as one that refers still to actual embodied beings, not the phantom disembodied beings sometimes imagined and created by popular films such as *A.I. Artificial Intelligence* (Spielberg, 2001). Although I researched this book primarily via textual analyses, I have noted in several places that I am considering Google Earth as a new and different kind of pan-sensual interactive text. In line with this, I suggest that the testimonies that suffering people offer through their photographic and written re- presentations on activist websites need to be considered as what they are *per se*, not only through the trope of ethical dilemmas of vision, mediated or not. We interpret the texts for their content, their histories of making and reception but we also need to accept our role of witness. This role of

[55] N. Kathryne Hayles, *How We Became Posthuman* (Chicago, IL: University of Chicago Press, 1999), 246.

witnessing needs to happen even when we are still embracing debates about the ethics of vision. If we accept that the content of a simulated, serious digital text such as Google Earth aims to and does represent actuality, then there is an ethical imperative to address the people whose images are embedded in this text, as actual people. I thereby witness their existence. Accordingly, I find myself interested in how the re-presented body can look back at the viewer. As I noted earlier, there is a moral need to engage with the people and places that are *inside* the image if we are to take on the responsibility of looking at them.

Paul Willemen defines a 'look back at the viewer' as 'the fourth look' in cinema.[56] Willemen's idea extracts the body of the filmed person from the space of the screen to look at us the viewers with eyes that stare down the lens of the camera. The fourth look is when the people we see on the screen confront us with our act of looking at them, by looking back at us via the same camera through which we are able to see them. He extends his definition of this look to include the sense in which a whole filmic text can confront an audience with their act of looking at the screen and its content. This confrontation is achieved through reflexivity in story, sound, music, characterization, location, set design or camera style. In describing the active subjectivity of a person viewing a film, Vivian Sobchack notes how this confrontation takes place from the point of view of the spectator:

> From the perspective of the subject of vision, that body (the spectator's) is not passive or 'empty'. It is a lived-body, informed by its particular sensible experience and charged with its own intentional impetus.[57]

Such representational audio-visual practices that confront us as embodied spectators through the *framing* of image content, unsettle the hegemonic status of vision by offering the possibility for what Haraway calls 'situated knowledges'. These kinds of knowledge refer to bodies located in specified places; they support an understanding of our eyes as embodied organic tools of vision, and not as simply signifiers of, in Haraway's words, 'a perverse

---

[56] Paul Willemen and Meaghan Morris, *Looks and Frictions: Essays in Cultural Studies and Film Theory* (London: BFI Publishing and Indiana University Press, 1994), 107.

[57] Vivian Sobchack, *The Address of the Eye: A Phenomenology of Film Experience* (Princeton: Princeton University Press, 1992), 305.

capacity to distance the knowing subject from everybody and everything in the interests of unfettered power'.[58]

So when looking at 'Crisis in Darfur' with my powerful and 'naked eyes' (also via my capacity for interactive computer vision) I literally know through the technology of Google Earth the spatial and temporal coordinates of the people I see and hear. My knowledge is explicitly situated in space and time, and depending on my response to these images, my knowledge can deepen into what Nussbaum defines as compassion: 'a certain sort of reasoning', 'a certain sort of thought about the well-being of others'.[59] As does Butler's concept of 'the stranger', Nussbaum argues that pity/compassion towards suffering people is requisite to social justice. She argues that compassion is a form of affective, imaginative social knowledge that brings the spectator into an engagement with the suffering person – a knowledge that is based on a sense of what human well-being is. This knowledge should encompass an awareness that suffering can come to us as well. She writes:

> The good of others means nothing to us in the abstract or antecedently. Only when it is brought into relation with that which we already understand – with our intense love of a parent, our passionate need for comfort and security – does such a thing start to matter deeply.[60]

I suggest that Google Earth offers an opportunity to extend the compassionate correlation between what we already understand and what is happening to people we do not personally know. This opportunity is through a simple device achieved via the complex, interactive technologies of Google Earth: we can tag or otherwise mark our personal domestic times and spaces on the same animated space as we find the marked spaces of others. We are performing inside the same *social space* as those with whom we engage in 'Crisis in Darfur'.

As noted earlier, Stahl focuses his investigation into how Google Earth acts 'as a kind of text, a powerful public screen onto which a political landscape is projected and thereby made sensible'.[61] Our growing intelligence of

---

[58] Haraway, *Simians and Cyborgs*, 188.
[59] Nussbaum, 'Compassion', 28.
[60] Nussbaum, 'Compassion', 48.
[61] Stahl, 'Becoming Bombs', 67.

interactivity and subsequent different levels of immersion contribute to how we interpret such explicitly interactive texts as Google Earth. Another way to understand our interactive, participatory use of Google Earth is to consider our participation as occurring within the fraught, ambiguously determined domain of how we might be said to 'perform' Google Earth. And if we employ the trope of performance to our textual interpretations then we must confront the possibilities of feeling and emotion, of affect, of the knowledge that these interpretations might bring us.

An augmented reality – Performing with our naked eyes

Haraway asks us to consider our collaboration with a machine as a collaboration with an 'other':

> All these pictures of the world should not be allegories of infinite mobility and interchangeability, but of elaborate specificity and difference and the loving care people might take to learn how to see faithfully from another's point of view, even when the other is our own machine.[62]

In this sense, we can think of ourselves as in a performative partnership with a machine that also sees. Our specific uses of Google Earth can be thought of as a dance between what we can see and touch with our eyes and what the machine can see and touch and then between what we can both see together as a consequence.

Our performance of engagement with images of suffering most often involves to some degree the experience of fear and of SHOCK. In human rights sites such as 'Crisis in Darfur' we confront shocking reportage and spectacle, the contextualization of such images and words with each other and also our contact with a massive amount of representation in the context of other genocides and suffering, together with all the associated sober discourse, as well as our memories and knowledge that we have gained from other sources.

---

[62] Haraway, *Simians, Cyborgs and Women*, 190.

When entering into a website such as 'Crisis in Darfur' we are not opening a book, a newspaper or a computer, or watching a television. We are looking for information within a new kind of context, one which we need to make for ourselves through our navigation of ever-changing website design and content. We are tracking these horrors and people via a personalized surveillance technology. Yes, we can use what we find for other (pitiless) purposes, but because the images, sounds and stories are open to different kinds of interpretation, it cannot be ruled out that one kind of interpretation might consist of the act of pity, of compassion, which 'is, above all, a certain sort of thought about the well-being of others ... a certain sort of reasoning.'[63]

When I see the flames of 'Crisis in Darfur' growing larger on my computer screen as I roll my mouse towards a closer focus, I feel a sense of dread and fascination. Why fascination? Perhaps it is not only towards a spectacle of destruction (as in the experience of car accident gazing) in which happily I am not directly involved. Perhaps such a fascination and dread is also a result of the recognition that Nussbaum names 'sympathetic identification.'[64] In other words, I know I am looking at pain, that I will be looking at people in pain if I move any closer, and I might allow or be surprised into allowing myself to feel the emotional pain of compassion. The knowledge that comes from a perceptual pairing with another is inherent to what I see and hear on this eclectic site. Their pain might have been mine if I was in the same circumstances. Compassion is affective, emotional knowledge that cancels out Descartes' 'mind/body split', and we are left again with bodies, theirs and ours. These bodies are situated in both our life stories and theirs.

We scramble after these connections between our lives and others' lives. These connections that happen in specific places and times are not obvious; we need to work at them in order to reach the position of compassionate knowledge. We cannot reduce this kind of HCI-mediated engagement to that which might be described by television theory's 'glance' or even through the

[63] Nussbaum, 'Compassion', 28.
[64] Nussbaum, 'Compassion', 39.

Dionysian gaze of the wandering *flâneur* of Baudelaire and Benjamin.[65] As new media theorist Tara McPherson suggests:

> We move from the glance-or-gaze that theorists have named as our primary engagements with television (or film) toward the scan-and-search ... a fear of missing the next experience or the next piece of data.[66]

McPherson goes on to note that our engagement with the web 'is not just channel-surfing: it feels like we're wedding space and time... The scan-and-search feels more active than the glance-or-gaze'.[67] As discussed earlier, this particular 'feel' is an outcome, at least in part, of interactivity at the HCI. In the immersive space of 'Crisis in Darfur', I can interact with the 'far away' in time and space. All this through the satellite technologies of remote sensing which the artist Caroline Bassett claims 'make it possible to touch a surface, to interrogate it, without being in direct contact with it. This is touch at a distance...'.[68] Bassett goes on to further describe the possibilities of remote sensing in the context of her artistic practice:

> Remote sensing thus suggests profound transformations in human sense perception, part of a broader series of (technologically influenced) shifts that are having an impact not only on scientific processes, but also on everyday life.[69]

Through remote sensing then, we find a new way of disappearing far distance; we are enabled in a new way to experience Nussbaum's 'thought experiment of compassion'.[70]

Can we understand the performance of this kind of embodied knowledge that is compassion, as an activist action of itself? Can we understand the Google Earth Outreach as a programme of activism? Is 'Crisis in Darfur' a human rights activist site? These questions do not relate to an issue of 'soft

---

[65] See again Kingsbury and Jones, 'Walter Benjamin's Dionysian Adventures'.
[66] Tara McPherson, 'Reload: Liveness, Mobility, and the Web', in *New Media, Old Media: A History and Theory Reader*, eds. Wendy Hui Kyong Chun and Thomas Keenan (New York, NY: Routledge, 2006), 204.
[67] McPherson, *New Media, Old Media: A History and Theory Reader*, 204.
[68] Bassett, 'Remote Sensing', 200.
[69] Bassett, 'Remote Sensing'.
[70] Nussbaum, 'Compassion', 52.

armchair activism' and voyeurism; they raise an inquiry into whether or not a certain kind of affective experience, derived from an interactive aesthetic, can be described as one that is socially useful and, to a degree, necessary for political action. Answering questions about how effective sites such as 'Crisis in Darfur' can be as a form of activism, requires a return to the definition of compassion as an active state of 'knowledge [which] is based on embodied subjectivity and that this form of knowledge is action'.[71] The following quote is from Kathy Marmor's discussion of Steven Holloway's 2005 performance 'One Pixel: An Act of Kindness':

> If knowledge is the capacity for action, then there must also exist power ... When these two conditions are met, then there exists the possibility of agency.[72]

Marmor's words are important in relation to thinking about engaging with the aesthetic of Google Earth, with both its offerings of knowledge enhanced through the sensation of flight and the web pages that Google Earth Outreach makes available through its Global Awareness Layers. Through Google Earth's aesthetic, there is an inherent sense of power/agency and a fairly immediate sense of acting out this power at an individual level.

Yes, we can pick up a pen or Visa card and donate to one of the organizations contributing to these Global Awareness Layers – this is one of the more overt manifestations of agency in relation to the knowledge acquired from accessing these and related sites. However, I suggest there is another primary agency involved with the actual acquisition of such knowledge. I suggest that our investigative engagement with Google Earth can actually constitute political engagement. It is a moot point whether or not we call this political engagement activism or the kind of awareness that is the precursor of more conventionally understood acts of activism.

The morality of this engagement is another matter. While I will not be discussing, in any detail, the actual possibilities (or lack of them) regarding specific capacities for action, I do suggest that by understanding

---

[71] Kathy Marmor, 'Bird Watching: An Introduction to Amateur Satellite Spotting', *Leonardo* 41, no. 4 (2008): 322.
[72] Marmor, 'Bird Watching', 323.

the practice of compassion as an active kind of knowledge, we can also describe a particular kind of knowledge-based agency that can be an outcome of our personalized journeys in Google Earth.

'Crisis in Darfur' offers access to knowledge via its presentation of stories, witness statements, photographs, statistics and videos that are laid over/embedded in a vast topography of human destruction. Some icons introduce us to higher resolution shots of Sudan, zooming across landscapes of burnt villages and refugee tent cities. These 'close-ups' of other people, other places and situations beyond our own live experience confront us with the perceptual challenges inherent to the era of globalized and digital technologies – a confrontation that we experience whenever we look out a plane window on to the landscapes of the Earth beneath us or when we fly through the cyberspace images from virtual globes like Google Earth.

5

Human Rights Activism via Remote Sensing – An Extended Case Study of Activism in 'Crisis in Darfur'

While not discounting the role of software in ordering the presentation of spatial data... we seek to understand how connections between place and cyberspace, what we term cyberscape, are continually (re)made.

Michael Crutcher and Matthew Zook[1]

In this chapter, I focus again but more deeply into the human rights site 'Crisis in Darfur' to tease out further how it manifests as a 'cyberscape' that operates within the practice domain of online activism. Firstly, I present a discussion of activism that begins with considering discourses of the social good and the public sphere. I then locate online activism within the more general domain of offline activism. Both the literature and practice of online activism are now extensive and varied, and the relationship between offline activism and online activism within the virtual worlds of communication and awareness raising is very complex. The one bleeds into the other and increasingly they are codependent. As discussed earlier in this book, the world of the actual now includes an acknowledgement that the virtual occurs within the wide parameters of the actual; we frequently act on what we know and imagine in virtual worlds (although more in some than others). In other words, digitally created virtual worlds have consequences in the actual world.

---

[1] Crutcher and Zook, *Geoforum*, 524. doi: 10.1016/j.geoforum.2009.01.003.

In this chapter I address the question: when and how can virtual worlds be instances of activism? It is interesting that the term online activism has such firm grounding in the literature of digital media, but how do we *act* when we participate in online activism? In the last chapter, I suggested that the Google Outreach programme could itself be considered as activist behaviour, although it is usually not understood as such. Indeed Rebecca Moore, founder of Google Earth Outreach, rather describes the Outreach programme as *support* for activist behaviour primarily through raising awareness of issues through offering and showing complex topographical images of conflict areas, be they directed at human rights or environmental issues.[2]

In the same context, however, that I think Google Earth might in Brian Massumi's sense be thought of as an 'event' – a specific realm of affect – I suggest that it is also possible to understand Google Earth Outreach to be an instrument of activism as well as an instrument that supports activism. This does expose Google Earth Outreach to implications of political action, and political action in turn is always vulnerable to accusations of bias and sometimes even to accusations of contributing to threats towards individual bodily safety, and national integrity and security.[3] Processes of human rights and environmental activism nevertheless hold a strong and growing moral imperative in global public discourse, although more so in developed, privileged societies.

My discussion here focuses on how the site 'Crisis in Darfur' can be thought of *as* activism: as a site both *for* and *of* human rights activism. In the previous chapter, I looked at how we might gain compassionate knowledge of other people through our responses to the site, via our finding and viewing of images embedded in this Google Outreach layer. In this chapter, I look particularly at one still image and one moving image in the context of a selection of other non-fiction images or documentary moving image sequences: all appear on the web and can be found in the 'Crisis in Darfur' layer. The documentary tropes used in these case studies are of photojournalism and documentary journalism. The still image is a photograph of two children in a refugee camp,

---

[2] GEOE Interview, 14 February 2014.

[3] For example, the current international diplomatic and security problems that swirl around the documents leaked in WikiLeaks and by Edward Snowden.

which is embedded in the layer; moving image sequences show children's drawings of traumatic events that happened to them.

My discussion then travels towards various other websites that document and comment on the Darfur/Sudan conflict both currently and during the primary active period of the 'Crisis in Darfur' layer, 2007–2009. These sites include clips shown on the distribution site YouTube, the Satellite Sentinel Project (SSP), stand-alone websites of the United States Holocaust Memorial Museum (USHMM), Human Rights Watch and Amnesty International. In the second section, 'Web Weaving', I set up the two theoretic tropes to describe the affective haunting that occurs via documentary web images: the idea of *ecos* and Bertold Brecht's concepts of *gest* and *social gest*. Using these tropes, I also examine the case studies as examples of new documentary gestures that are emerging on the World Wide Web (WWW). Drawing on Jon Dovey's and Mandy Rose's words, one of the aims of this chapter is to 'work within the memory of documentary as social praxis in its attempt to argue for new modalities of coherence within the emergent online environment'.[4] To establish the online context of my case studies, it is useful first to consider online activism as the communication media practice in which Google Earth is implicated at the 'Crisis in Darfur' layer.

Online activism: Working for the social good in the public sphere of the web

In their introduction to their edited volume *Communities of Sense*, Beth Hinderliter et al. write the following about the relationship between politics and aesthetics: '[A]esthetics is not taken as grounds for, but as a means to construct the possibility of, shared meaning.'[5] This very simple statement contains a complex and much debated thesis. It introduces the idea that a set of aesthetics need not simply be understood to constrain and thereby define

---

[4] Jon Dovey and Mandy Rose, 'We're Happy and We Know It: Documentary, Data, Montage', *Studies in Documentary Film* 6, no. 2 (2012): 160.

[5] Beth Hinderliter et al., 'Introduction', in *Communities of Sense. Rethinking Aesthetics and Politics*, eds Beth Hinderliter et al. (Durham: Duke University Press, 2009), 19.

meaning, but that the aesthetics of a communicative medium also can offer an interpretive and experiential space where people converse, negotiate and describe their world. This discursive, public space is explicitly interactive when it happens online – it can only be created by people interacting with each other in an online space. In order to examine the social worth of this kind of interaction, it is worth interrogating how this online space might overlap with Jurgen Habermas' idea of 'the public sphere' – a communicative space where a particular kind of civic engagement occurs.[6]

This quote from Zizi Papacharissi[7] well defines how the public sphere can currently be understood:

> The public sphere presents a domain of social life in which public opinion is expressed by means of rational public discourse and debate. The ultimate goal of the public sphere is public accord and decision-making, although these goals may not necessarily routinely be achieved … the value of the public sphere lies in its ability to facilitate uninhibited and diverse discussion of public affairs, thus typifying democratic traditions.[8]

He says that this rendition of the public sphere should not, however, be taken simply to mean 'public space' and that nevertheless the public space of cyberspace does offer another kind of public sphere: a mediated public sphere. I suggest that within the liberal domain of this digitally mediated public sphere, online/Internet activism can happen.[9] Papacharissi's discussion of the public sphere that exists via the Internet and WWW eloquently acknowledges the difficulties in reconciling the rationality of Habermas' idea with the often banal and poorly informed opinions that are a part of the blogosphere. He interrogates the much and often debated borders between private and public communication in the age of digital social media, and develops an

---

6   Jurgen Habermas, *Theory and Practice*, trans. J. Viertel (London: Heinemann, 1973), 351 as quoted in Zizi Papacharissi, 'The Virtual Sphere 2.0. The Internet, the Public Sphere, and Beyond', in *The Routledge Handbook of Internet Politics*, eds Andrew Chadwick and Philip N. Howard (London: Routledge, 2009), 232.

7   Zizi Papacharissi is a Professor and Head of the Department of Communication at the University of Illinois, Chicago, USA.

8   Papacharissi, 'The Virtual Sphere 2.0', 232.

9   Reflecting its usefulness, this idea of Internet space as public sphere is used as self-evident by Jon Dovey in his essay 'Simulating the Public Sphere', in *Rethinking Documentary: New Perspectives and Practices*, eds T. Austin and W. de Jong (Maidenhead: McGraw Hill, 2008), 246–257. ISBN 978-0335221912.

argument that uses the work of political philosophers Hannah Arendt[10] and Chantal Mouffe[11] to locate and describe a new kind of discursive sphere where individuals can work towards the social good of a society. Papacharissi describes the new digitally enabled private sphere as 'empowering, liquid, and reflexive'.[12] He argues that it is the 'reflexive architecture' of digital *online* media that now creates the connection that is much needed for political action, between civic-minded individuals and their societies; and that this digitally enabled connectivity thereby cancels out the disempowering isolation that Arendt feared could result 'when all political action retreats to the private sphere'.[13]

The vast breadth of the WWW as enabling a public sphere also means that hegemonies of discourse emerge on a large scale; and within this mediated public sphere, debates can be censored for political and economic reasons as well as culturally based moral judgements. Whilst every censor might argue that a particular case of censorship is for the public or social good, and whilst this mediated public sphere also contains the sublime, the ridiculous and the poorly thought-through opinions and theories of massive amounts of people, the WWW is nevertheless a public sphere that offers much in the way of public good. Online communities have the potential for both the exposition of ideas, opinions and judgements of the many, with the exposure of those who have them and how they have evolved. The absence and presence of voices also can expose those ideas, activities and people that various (corporate, state and institutional) hegemonies perceive as threats to their power arrangements and agenda.

The digitally mediated public sphere that incorporates Internet/online activism uses a range of communicative activities offered within a variety of aesthetics. These include blogs, websites, Twitter feeds, news feeds, YouTube and Facebook pages that give information about activist offline engagements happening in real time, or those that are planned for the future. These web

---

[10] See Hannah Arendt, *Between Past and Future: Eight Exercises in Political Thought* (New York, NY: Viking, 1968), 4, quoted in Papacharissi, 'The Virtual Sphere 2.0', 244.

[11] See Chantal Mouffe, *On the Political* (London: Routledge, 2005) as used in Papacharissi, 'The Virtual Sphere 2.0', 241.

[12] Papacharissi, 'The Virtual Sphere 2.0', 244.

[13] Papacharissi, 'The Virtual Sphere 2.0', 244.

pages can also act as archives of past activities and as sources of information. To quote human rights media theorist Leshu Torchin:

> Broadly speaking, Internet activism refers to the use of online technologies such as e-mail, file sharing, and websites to organize communities, raise funds, and lobby for change.[14]

Torchin goes on to note how the Internet was recognized as a powerful new medium for advocacy before the advent of Web 2.0 and describes the emergence of huge interactive social media sites from 2004 onwards.[15] One site that has developed a social media platform for citizen journalism and advocacy is the social media site YouTube. In 2008, YouTube launched CitizenTube, a moderated blog space for posting videos in advocacy campaigns and for 'breaking news' items posted by both citizen and professional journalists. It now uses Twitter feeds as well.

The earlier, pre-2004 kind of online activism relied explicitly on how the Internet could be used for situating people in communities which existed in specific online spaces directly affiliated to specific offline places: 'DigiPlace'. It is worth unravelling some of the connotations of this term because it neatly sums up a number of problems inherent to collisions and convergences between the virtual and the actual/live: mediated space. Michael Crutcher and Matthew Zook, in their article 'Placemarks and waterlines: Racialized cyberscapes in post-Katrina Google Earth', describe how DigiPlace is created by the power of software (via the power of those who know how to make software) and that it 'represents the interface to the blending of the material and the digital'.[16] Their discussion of DigiPlace[17] focuses on how search engine algorithms make some people, geographies and social issues visible and keep others invisible, or on the periphery of online visibility. DigiPlace is a term then that also can be

---

[14] Torchin, *Creating the Witness*, 173.

[15] For an overview of Internet activism prior to 2001, Torchin specifically refers to Martha McCaughey and Michael D. Ayers, eds, *Cyber-activism: Online Activism in Theory and Practice* (New York, NY: Routledge, 2003). Another more recent (Web 2.0 and onwards) collection of essays on new forms of media-enabled activism is Susan Curry Jansen, Jefferson Pooley and Lora Taub-Pervizpour, eds, *Media and Social Justice* (New York, NY: Palgrave Macmillan, 2011).

[16] Crutcher and Zook, 'Placemarks and Waterlines', 524.

[17] Crutcher's and Zook's discussion of DigiPlace draws on an earlier article written by Zook with Mark Graham, 'The Creative Reconstruction of the Internet: Google and the Privatization of Cyberspace and DigiPlace', *Geoforum* 38, no. 6 (2007): 1322–1343.

used to describe and interrogate power relations which come about through both the presence and the absence of some points of view in an online activist intervention. Where some points of view are made clear whilst others are ignored in order to make the case for a particular intervention.

Although DigiPlace is a term developed to refer to inequalities between people, places and issues that are highly visible on the WWW and those that are not, the reasons for such inequalities are not necessarily born of ill intentions and sometimes emerge inadvertently from a perceived need for rhetorical force. An example of DigiPlace that has had a high profile in the first decade of the 2000s is citizen online journalism. This particular form of DigiPlace includes all the things that have happened under its auspices for both the social good and the detrimental visibility of some people engaged in online activism, as well as the abysmal invisibility of others who do not attract the attention of amateur or professional journalists. If you can't be seen or noticed on the WWW then your interests will not be seen or noticed either: the WWW is the last resort for attracting attention.

After the emergence of big social media sites, activism has continued its activities in DigiPlace. I think, however, that it is fair to say that these big social media sites have also provided 'a cloak of respectability', a new sense of authority to people involved in online activism. One reason for this is because they offer a very large conduit for distribution: for example, if you have a lot of 'likes' on your site then people assume that their agenda is also that of a lot of other people and that it follows that they now have the status of authority about whatever it is they are saying or showing on their website, and that this shared information is thereby validated as 'news'. Activist sites, including those using Facebook, Twitter and Instagram, now offer information as 'news items' primarily in the form of written text conversation, photographs and with increasing bandwidth now more available, short audiovisual clips. The challenges that these sites of 'alternative journalism' (better known as 'citizen journalism') present to professional journalism have been well discussed and debated.[18] The

---

[18] For example of a study on online citizen journalism and its offline counterpart as well, see Luke Goode, 'Social News, Citizen Journalism and Democracy', *New Media and Society* 11 (2009): 1287–1305. doi: 10.1177/1461444809341393.

rise of 'citizen journalism' nevertheless continues, although some of its consequences for individual people have emerged relatively slowly, for example, the escalating risks to personal security of all kinds of journalists, both professional and amateur. Indeed with the huge reach of the WWW and mobile digital devices such as digital camera phones and personal computers, any journalistic mistakes can have consequences that go far beyond declarations of what information is verifiable or not.

The array of revolutions and uprisings occurring since 2010 in North African countries and the Middle East has become the most well-known example of using digital media to disseminate information linking groups of on-the-ground activists with each other and with communities around the world. This set of revolutions, civil war and other political uprisings began in Tunisia in December 2010 and continue. They include the uprisings that have resulted in successful attempts (Tunisia, and Egypt twice) and so far unsuccessful attempts (Syria) to oust national leaders seen to be corrupt, incompetent and unrepresentative of the majority or of large minorities in the population. In 2011, the autocratic governments of Tunisia, Libya, Egypt and Yemen all fell, and in the context of these social uprisings, digital media has been seen as a tool for democratization.

Major research has been done on how effective both offline and online activism has been in these uprisings, particularly in the cases of Tunisia and Egypt. For example, the site called 'Journalist's Resource' is a research portal and curated database that refers to a number of articles and studies on this topic, which have been published in peer-refereed journals as well as in magazines and other news media.[19] Another example of major research into the conjunction between social unrest and digital media is a report from the United States Institute of Peace, even before The Arab Spring of 2011. This report allocates five levels of analysis for examining the impact of new media on collective community actions, especially focusing on the social media sites Twitter, Facebook and YouTube during social uprisings. The following is a quote from this report:

---

[19] 'Journalist's Resource', The Shorenstein Center on Media, Politics and Public Policy, Harvard Kennedy School, http://journalistsresource.org/

New media have the potential to change how citizens think or act, mitigate or exacerbate group conflict, facilitate collective action, spur a backlash among regimes, and garner international attention toward a given country.[20]

The report directs attention to the 'Green Revolution' that occurred during and just after the 2009 elections in Iran. In its summary, the report delivers an early warning on how citizen journalism and online activism can result in risks to the personal safety of activists and how the same digital media used by activists can also be used by the regimes they are challenging:

> Although there is reason to believe the Iranian case exposes the potential benefits of new media, other evidence – such as the Iranian regime's use of the same social network tools to harass, identify, and imprison protesters – suggests that, like any media, the Internet is not a 'magic bullet'. At best, it may be a 'rusty bullet'. Indeed, it is plausible that traditional media sources were equally if not more important.[21]

A clear and distressing example of how social media can be used badly and yet mostly with the best of intentions is the case of a young woman called Neda Agha-Soltan who was shot in the neck and died during the Iranian demonstrations. Her death was captured by people around her who were using camera phones; the footage was then quickly posted on YouTube. The video footage shows one of the most disturbing images produced within this kind of reportage activism. We watch the young woman being attended to by a doctor who was also in the crowd. With her eyes watching us, we watch her die.[22] A huge outcry followed this death, and a movement emerged in Iran called 'I am Neda'. This involved demonstrations by crowds holding over their faces

---

[20] Summary of Sean Aday et al. *Blogs and Bullets: New Media in Contentious Politics*, Report from the United States Institute of Peace's Centers of Innovation for Science, Technology, and Peacebuilding, and Media, Conflict, and Peacebuilding, a team of scholars from The George Washington University, in cooperation with scholars from Harvard University and Morningside Analytics, critically assesses both the 'cyberutopian' and 'cybersceptic' perspectives on the impact of new media on political movements, published on September 2010. The link to this summary is http://www.usip.org/publications/blogs-and-bullets-new-media-in-contentious-politics (accessed 3 March 2014).

[21] Aday et al., *Blogs and Bullets*.

[22] For background to this incident, see the YouTube clip http://www.youtube.com/watch?v=wXN_yCSbUYk (accessed 3 March 2014), and for the clip of Neda's death contextualized with commentary, see http://www.youtube.com/watch?v=o-jXwi_QEgI (accessed 3 March 2014).

masks of Neda. Unfortunately, the image of Neda was taken from a search on a social media site and was not of the young woman who died but of a university lecturer called Neda Soltani who was subsequently harassed by the Iranian government at the time.[23]

The pre-Web 2.0 kind of Internet activism that Torchin broadly defines and the activist use of social media in political uprisings in live space are only two kinds of activity employed by activists who use the Internet and the WWW. Another is better known as 'hactivism' and its very definition is contested by hactivists themselves.

Rita Raley[24] writes on the history of the term 'hactivist' and how it has been understood within the domain of online activism to infer both affirmative and subversive coding actions: an aggressive weapon of assault by interference via '[d]istributed denial-of-service attacks'.[25] On the other hand, Raley also refers to Oxblood Ruffin of the hacking collective Cult of the Dead Cow that first used the term hactivism and says that 'according to Ruffin hactivist networks are the "blue helmets" of the Internet and thereby ought to work toward open code and peace rather than war'.[26] Raley notes Ruffin's version of hactivism should work against net censorship rather than deny 'free speech and a violation of the principle of free flow'.[27]

It is important to note, however, that hactivism is reserved for those who know how to code, how to alter software programs and how to break through the protective firewalls of websites owned by nation states and corporations. Hactivism works within and is therefore part of the power relations that exist within the discourse of software creation. In Foucault's sense, hactivists subvert these institutional discourses because they can only exist within these discourses. An interesting development in the context of hactivism is the idea that hacking can constitute a tool for a new kind of warfare that is now called cyberwar.

[23] See Tracy McVeigh, 'Iranian Fugitive: Identity Mix-up with Shot Neda Wrecked My Life', *The Observer*, 14 October 2012, as posted on http://www.theguardian.com/world/2012/oct/14/iran-neda-soltani-id-mix-up (accessed 3 March 2014).

[24] Tactical media theorist Rita Raley is an Associate Professor in English at the University of California, Santa Barbara.

[25] Rita Raley, *Tactical Media* (Minneapolis, MN: University of Minnesota Press, 2009), 41.

[26] Raley, *Tactical Media*, 41.

[27] Raley, *Tactical Media*, 41.

Another example of disruptive actions that can and has been identified as a tool for cyberwar is 'swarming'. In the context of social justice activism, the word 'swarm' emerged from the acronym of the name for a particular intervention that took place on 1 May 2006: SWARM – South West Action to Resist the Minutemen. On that date, a collaboration between the Electronic Disturbance Theater and other activists from the Tijuana-San Diego area performed what Raley describes as a 'virtual sit-in'. This intervention was a denial-of-service action via a FloodNet application on the websites of paramilitary vigilante organizations that took upon themselves to monitor border crossings between Mexico and the United States.

Major hactivist organizations often have drawn on live theatre traditions to name themselves and their interventions, recalling the aims and conventions of Brechtian *agitprop* theatre.[28] Important seminal hactivist groups include Electronic Disturbance Theatre[29] (referred to earlier), the Critical Art Ensemble and the Tactical Media Crew.[30] The live/hactivist grouping that calls itself Anonymous is one of the most recent iterations of hactivism. Their credo is: 'We are Legion. We do not forgive. We do not forget. Expect us' and in accord with the aims of open distribution activism, the documentary film made about them is available on YouTube in its entirety: *We Are Legion: The story of the hactivists* (Brian Knappenberger 2012), http://www.youtube.com/watch?v=lSqurTMe7Rw.

Hactivism then can be thought of as a kind of insurgency action, with all the possibilities for social good and social harm that are the outcomes of insurgency actions. In this context, Raley tells us about Michael Dartnell's distinction between what he regards as hacking as 'denial of service' and Web activism, which he says does not so much interfere with the infrastructure of

---

[28] A useful definition of 'agitprop' in the context of hactivism can be found in Merriam-Webster online dictionary: 'Political strategy in which techniques of agitation and propaganda are used to influence public opinion. Originally described by the Marxist theorist *Georgy Plekhanov* and then by *Vladimir Ilich Lenin*, it called for both emotional and reasoned arguments.' http://www.merriam-webster.com/dictionary/agitprop (accessed 2 March 2014).

[29] For a clear discussion of the activities of the Electronic Disturbance Theatre and the future of hactivism, see Ricardo Dominguez, 'Electronic Civil Disobedience: Inventing the Future of Online Agitprop Theater', *PMLA* 124, no. 5 (2009): 1806–1812.

[30] For a useful account of hactivism as it emerged in the United States, see all of Raley's Chapter 1. 'Border Hacks: Electronic Civil Disobedience and the Politics of Immigration', in *Tactical Media*, 31–64.

a website but rather productively 'aims to transform the social conditions in which that infrastructure is situated'.[31]

Another kind of online insurgency is the subversion of surveillance technologies. A good example of this lies in Steve Mann's concept of 'sousveillance': 'watchful monitoring from below' rather than from 'above', a practice of inverse surveillance where surveillance technologies can be used to subvert the discursive regimes that produce them.[32] Visual artists have been at the forefront of 're-purposing' surveillance technologies. Recalling how the term DigiPlace can be used to interrogate power relations involved in being visible (or not) on the WWW, Andrea Mubi Brighenti interprets surveillance art 'as an attempt to deal with issues of social visibility and invisibility and, more specifically, with *visibility regimes*'.[33]

Artists have long been interested in using new technologies as they emerge and are often the first to imagine the possibilities of new technologies for both their artistic expression and their political rhetorical power. A complex and important example of this kind of activism is the online and offline activities of the organization WITNESS.org. This human rights activist organization works with other activist organizations or smaller groups to assist them with video equipment and training for advocacy campaigns.[34] It also organizes screenings of some videos that are deemed useful for lobbying decision makers in governments and corporations. It distributes and exhibits some videos on its own website.[35] This organization is a powerful promoter of sousveillance and draws on Torchin's understanding of the media's capability for bearing witness to conflict, injustice and trauma: that 'audiences can become witnesses through watching'.[36]

Another activism format is 'persuasive gaming' or what is also known as 'serious gaming'. This type of activism has been used by activist organizations, militaries and militant organizations of many political

[31] Raley, *Tactical Media*, 42.
[32] As described by Mark Andrejevic, 'Watching Back. Surveillance as Activism', in *Media and Social Justice*, eds Sue Curry Jansen et al. (New York, NY: Palgrave, 2011), 179.
[33] Andrea Mubi Brighenti, 'Artveillance: At the Crossroads of Art and Surveillance', *Surveillance and Society* 7, no. 2 (2010): 176. ISSN 1477-7487.
[34] For a clear discussion of WITNESS.org, see Torchin, *Creating the Witness*, Chapter 4, 136–171.
[35] Torchin, *Creating the Witness*, 148.
[36] Torchin, *Creating the Witness*, 7.

affiliations, including the US military and al-Qaeda. According to Raley, game modifications and custom-built simulations can offer 'sympathy and identification' in the same way as 'cinematic imagination invites a certain cathexis from the viewer, as does the first-person point of view. Representation, or in this case simulation, paves the way for real experience.'[37] In the context of 'Crisis in Darfur', one game in particular stands out for how it draws attention to the war in Darfur: 'Darfur Is Dying'. This game was initiated by Games for Change and launched at a 'Save Darfur' rally in Washington, D.C., in 2006.[38] Developed primarily by media artist and scholar Susana Ruiz, it positions the player as a refugee in one of the huge camps for refugees from Darfur and offers an interactive perspective on what it is like to live in the camps.[39]

In the context of online insurgency and other kinds of online activism and witnessing, I conclude then that Google Earth's visualization of Geographical Information Systems (GIS) data can be used for exercises in sousveillance – for using the satellite remote sensing technology that was initially and still is used by militaries around the world, to subvert the hegemony of visibility that is prevalent in surveillance operations. It is also possible to conclude that Google Earth draws much of its effectiveness for witnessing environmental catastrophes and human rights violations, from its 'game-like' first-person positioning of its users and participants. In other words, this part of Google Earth's aesthetic contributes to its activist function.

An aesthetic of the possible

Returning to how the aesthetics of a communicative medium can both enhance and constrain information content, it is possible to distinguish between Google Earth and other vehicles/media for online activism. Google Earth is a text that uses a specific set of aesthetics: interactivity, movement,

---

[37] Raley, *Tactical Media*, 78.
[38] For background and information on this game, see the web page http://www.darfurisdying.com/aboutgame.html (accessed 13 March 2014).
[39] Also see Torchin, *Creating the Witness*, 193.

the simulated flight and the first-person point of view of first-person shooter computer games, realistic animation, saturated colour and written text. One of its most overwhelming aesthetic components, however, is an ephemeral yet considerable sense of possibility: of the potential for looking, of understanding the world as simultaneously both far away and 'up close and personal'. This sense of the possible is different from the more directed, focused speculative imaginings that are the outcomes of simulation and other modelling processes. A sense of the possible, an awareness of new possibilities for information and interpretation also distinguishes online activism that searches for images that come via remote sensing. That sometimes frantic and sometimes delightful feeling of looking for something you need/want to know about or even the leisurely feeling of 'just looking' is one of the experiences that search engines can provide us with. Google Earth certainly can provide us with this experience. Google Earth also visualizes for us the space of our searching. It visualizes a specific field of search for the possible: visions of Earth from the point of view of satellites orbiting the Earth. One of Google Earth's major current projects is to present to its users and participants, the experience of gaining information as a 'unified web'.[40] One of the ways in which facility is made available is through the app GeoFind:

> GeoFind extends Google Earth by integrating the Google Search Appliance into a seamless user experience. This brings together Google's powerful search technology and Earth's geospatial analysis and visualization capabilities to provide instant, relevant, geocoded results.[41]

Another way in which this unified web will be made possible is through an easily accessible index of KML layers. To an extent, and as noted in Chapter 3, this is already available via the Google Earth Gallery, but now it will be searchable, with Google Earth showcasing activist layers that it judges merit further exposure. This offering from Google Earth will create some much needed order into the enormous array of layers that exist already whilst also allowing people to conduct focused searches instead of having to rely on

---

[40] GEOE Interview, 14 February 2014.

[41] https://www.google.com/enterprise/marketplace/viewListing?productListingId=6239+1515056225
6131649073&pli=1 (accessed 1 March 2014).

a *flâneur* style of wandering through sites that have been already chosen, exhibited for our curious and/or purposeful gaze. So an approach that uses the idea of a unified web does not so much close down or confine our searching of Google Earth to visions of Earth, but opens up other ways to find GIS, other forms of social information and other ways of relating them to each other. Although Google is placing its search engines at the service of our searching, this also means the corporation simultaneously directs our searching via its ranking algorithms. Nevertheless, I argue here that Google Earth with all its apps and Outreach programmes works through an aesthetic that opens up pathways to information and knowledge by showing us (via animated modelling) an overall vision and interpretation of the Earth as whole, as if seen from space. The platform gives us the possibilities of an astronaut's vision, allowing us to care and yet not be burdened with the many ethical and practical responsibilities of that kind of vision. It offers a free, if mediated, trip around the world.

In their discussion of that amalgam of cyberspace and the actual which they define as DigiPlace, Crutcher and Zook note how '[t]he ability to simultaneous move through the offline and the online worlds... is shaping the human landscapes we inhabit.'[42] This point is highly relevant to all kinds of online activism, but is newly relevant when thinking about activism that stems from information taken from remote sensing satellites and used for the practices of digital cartography. Sites such as Google Earth with its layers overtly display how human landscapes are embedded in actual places on Earth. Since the GIS data visualized in Google Earth is obtained from remote sensing satellites orbiting the Earth, I think it is possible to say that Google Earth is an example of not only a site for online activism but specifically a site for remote sensing activism, a new kind of activism also taken up now by several activist organizations on their own websites as well as in the KML layers that some of them have created within Google Earth.

This new kind of activism deploys a new kind of witnessing, one that relies on remote sensing of landscapes, people and situations otherwise relegated to the technologies of writing, photography and film-making.

[42] Crutcher and Zook, 'Placemarks and Waterlines', 524.

These older formats are layered over remote sensing data but this overlaying process relies on depictions of the earth that are sourced from space satellites orbiting the Earth. In the next section of this chapter, I closely address specific images captured in the layer 'Crisis in Darfur' that form activist interventions, and how these images are linked to the broad domain of online and offline human rights activism.

Web Weaving

How can we articulate the consequences of our web searching and surfing for other people and places? Bruno Latour's idea of 'the fold' is useful here, as he describes how the juxtaposition of what happens locally onto a broader framework can be better thought of as a 'fold' that produces a more useful context for both the local and the broader situation:

> When you put some local site 'inside' a larger framework, you are forced to *jump*. There is now a yawning break between what encloses and what is enclosed, between the more local and the more global. What would happen if we forbade any breaking or tearing and allowed only bending, stretching, and squeezing? Could we then go *continuously* from the local interaction to the many delegating actors? The departure point and all the points recognized as its origin would now remain *side by side* and a connection, a fold would be made visible.[43]

When we surf the WWW we create fold after fold after fold. Through our own personal connecting of ideas and websites to each other, we find new ways of looking at and thinking about the world. The wider framework might be one website from which we begin our surfing, or it might be the WWW itself. These folds and the connections we create with our surfing constitute one of the most interesting aspects of the WWW as a whole and of Google Earth in particular. One of the important kinds of folds that we create when using Google Earth is that between the far and the near. Sam Gregory, from the human rights activist organization WITNESS.org (http://www.witness.org),

---

[43] Bruno Latour, *Reassembling the Social. An Introduction to Actor-Network Theory* (Oxford: Oxford University Press, 2007), 173–174.

notes that the overall action that eventuates in finding such sites as Google Earth's 'Crisis in Darfur' could well be described as one that we 'stumble upon',[44] something that can surprise and/or satisfy a search for something unknown. This act of stumbling upon information via the web also describes well the nature of 'surfing the web'. In the case of Google Earth, however, 'surfing' is achieved via a very particular aesthetic, one that induces a sense of hovering over a world which is both familiar and yet made strange. Russian Formalism's idea of 'making strange' – *oestranenie* – is useful for describing how Google Earth offers an embodied sense of combined visual pleasure and weightlessness in flight, and in Fredric Jameson's words, through 'a renewal of perception'.[45] How then can we describe these folds, juxtapositions of near and far, familiar and strange that we come across when we search/surf the WWW? One way is to think about specific sites and images as belonging to an ecology of images and sites.

Ecos and media ecology

A metaphor for the holistic study of media use with particular reference to context and especially the interrelationships of a medium or tool with its users' tasks, roles, attitudes, and practices.

Oxford Reference Online

In line with this definition, rather than focusing only on the images as isolated animated entities with embedded meanings all on their own, I look here at their *ecos*: what makes up their 'household', the intimate community which they inhabit, drawing on the ancient Greek root of the word ecology, meaning 'house study'. My analysis thereby also focuses on some of the ever-changing web communities of images in which individual images exist.

The *ecos* of these images can of course also be described by the contexts of their creation, content and reception. But their description should include the spatial montage within which they are placed by all three

---

[44] Sam Gregory, personal communication, Visible Evidence Conference, Istanbul, 2010.
[45] Fredrik Jameson, *The Prison-House of Language: A Critical Account of Structuralism and Russian Formalism* (Princeton, NJ: Princeton University Press, 1972), 51.

contexts. Manovich's (2002) definition of 'spatial montage' – '[a] meaningful juxtaposition of more than one image stream within a single screen'[46] – is especially useful for describing how these images are placed in relation to others. Such juxtaposition can be within their 'domestic' community of sounds, texts and images (within a single screen space) as well as within a wider range of content to be found in linked screen spaces. These latter spaces depend on live links embedded in the original site and in the further extrapolating number of links resulting from these. While all of these sites might not be relevant at first glance, they, nevertheless, within the arena of one web surfing session, also contribute to the stories and the cultural meanings of images found within a human rights activist site. The potential cultural meanings available to one web surfing session can be the actual result from these images being placed side by side with others via their sequential or spatial placement over the vectors of time and mapping of geographical space as it becomes apparent in web space.

Such cultural meanings can occur on a small or large scale. They can result from the personal, quick as lightning flash of understanding that we can have of an image, drawing on our own history of viewing and wider experience. These are the kinds of understanding that I first apply in my analyses of the two case studies. Then there are the broader meanings that result from further reflection. Whilst also dependent on personal contexts of viewing, these meanings depend on the wider *ecos* of the images. I suggest that the process of how we become aware of cultural meanings from both the small and wider frames of viewing can to some extent be described through using two concepts from Bertold Brecht.

Theatre of the Web

I am not the first person to analyse a digital object via the trope of theatre. Leon Gurevitch specifically talks about the digital globe (e.g. Google Earth) as theatre and how this trope of theatre can 'challenge the idea that

---

[46] Lev Manovich, 'The Archaeology of Windows and Spatial Montage' (September 2002) http://www.manovich.net/DOCS/windows_montage.doc (accessed 15 January 2013).

it functions as a straightforward representation of data about the Earth'.[47] He goes on to suggest that 'the digital globe is already a place within which competing rhetoric is deployed by various "actors"'.[48] This is a valuable insight since it assigns agency, an assertion of the power of choice, to the people whose voices we hear and whose images we see on the WWW. For analysing my two case studies, I am using the trope of Brecht's 'epic theatre' to describe how these case studies are situated within 'Crisis in Darfur' and also within the broader and highly complex fabric of the WWW.

One of Brecht's most significant contributions to the practice and theory of theatre was his idea of *gest*: an 'overall attitude'[49] that he wants an actor to have towards a particular characterization or dramatic role. *Gest* is not mere 'gesticulation' although its practice does include tone of voice, body gestures and the way in which the actor moves and places her body in time and space. *Gest* is the actor's (and the director's) comment on what sort of person is being represented through a particular drama. Brecht further describes a *social gest*: 'the mimetic and gestural expression of the social relationships prevailing between people of a given period'.[50] These concepts of *gest* and *social gest* are useful in working out how images affect us; as I suggested in Chapter 4, they introduce a metaphor of theatricality that is significant when considering how the Web mediates and captures/documents/performs the actuality of people, places and moments in time. My application of these concepts in my analyses of the case studies describes visual (and audiovisual) texts as able to act in the world, to present actorly *gest*. My interpretation of *social gest* is applied to the whole experience of the creation, linking, viewing and manipulating these texts.

In Chapter 1, I noted Stahl's astute description of Google Earth as 'a powerful public screen onto which a political landscape is projected and thereby made sensible'.[51] The actual screen space on which we see the images generated in Google Earth is not necessarily large; it is usually the intimate

---

[47] Leon Gurevitch, 'The Digital Globe as Climatic Coming Attraction: From Theatrical Release to Theatre of War', *Canadian Journal of Communication* 38 (2013): 336.

[48] Gurevitch, 'The Digital Globe as Climatic Coming Attraction', 336.

[49] Bertold Brecht, *Brecht on Theatre. The Development of an Aesthetic*, ed. and trans. J. Willett (London: Eyre Methuen, 1978), 104.

[50] Brecht, *Brecht on Theatre*, 139.

[51] Stahl, 'Becoming Bombs', 67.

portal to the world that is our personal computer. What is interesting in relation to viewing moving and still images on the public screen of the Web is that they can easily be played over and over via the clicking of the 'mouse'. This repetitive kind of viewing exerts a form of performativity that includes the *gest* of the image text itself – how it depicts an 'overall attitude' to its social context. The *gest* of a screen image can of course be further affected by the websites and web pages that preceded and came afterwards in the journey of an individual's specific viewing of that image. Once images are on the Web they are hugely available and the producers of images have little control over the sequence of their viewing. The sociopolitical attitude, the *social gest*, is demonstrated both by our searching and linking of these images in our personalized surfing and by the often indirect ways they have been linked together via wider structural linking between websites that contain or refer to them. Discerning a *social gest* becomes important for working out and defining the potential direction of power and other social relations that are embedded within the images themselves. Power flows in both the intimate and broader contexts of their creation, content and reception and also importantly, their positioning over time.

As noted earlier, the 'actors' in my discussion are the images being examined here, not the subjects of the images. The *social gest* relevant to my discussion is the social situation that is being described by these images as understood by looking at the visual community in which they exist on the web. By juxtaposing my study of these specific images with an awareness of other photographs and videos produced by human rights activists, I am creating as any web surfer does, a particular community for them. Their *ecos* is one of war and genocide: photographs, videos, witness statements quoted in text, statistics of destruction all embedded in the veil of red and yellow icons of fire and burning that cover the site 'Crisis in Darfur'.

'Crisis in Darfur' – again

To backtrack a little to the history of the recent genocide in Darfur that is represented in 'Crisis in Darfur' together with its costs in lives and aid money,

I quote here from an article by Andrew S. Natsios who posted it on the Oxford University blog on 1 November 2011. Natsios served as Administrator of the U.S. Agency for International Development from 2001 to 2005 and was appointed as Special Humanitarian Coordinator for Sudan. He also served as Special Envoy to Sudan from October 2006 to December 2007. He is author of the 2012 monograph 'Sudan, South Sudan, and Darfur: What Everyone Needs to Know'.[52] The information quoted below is very much from a government-based aid worker's point of view:

> This long-neglected western region has been intermittently at war since the 1980s and claimed the lives of 300,000 Darfuris in its most recent phase. The rebellion beginning in 2002 led to an ongoing humanitarian emergency, costing Western governments about one billion dollars annually at the peak of the crisis to sustain the 1.8 million people driven into sixty-five IDP (internally displaced person) camps scattered across Darfur. The Sudanese government committed widespread atrocities in Darfur as part of its counter-insurgency strategy, which involved a massive ethnic cleansing campaign to displace the tribes that started the rebellion ... The crisis has led to the deployment of 26,000 United Nations/African Union (UN/AU) peacekeeping troops and police – the largest in UN history to a single conflict – to Darfur, which cost \$2 billion to maintain in 2007 alone.
>
> http://blog.oup.com/2011/11/poty-sudan/ (5 November 2011)

The particular stage of the conflict in Darfur that is documented in 'Crisis in Darfur' is known as the Third Darfur Rebellion. This conflict with its terrible massacres finally captured global attention in 2002. During the same period of time, the conflict between northern and southern Sudan also continued. Several and various rebel groups have emerged and still continue to emerge in both conflicts, with political alignment shifting and changing both inside these groups and between them. Although the nation of South Sudan was established in 2011 after much brokering by the United Nations, conflict is still current both between Sudan and South Sudan and within South Sudan itself. Natsios judges that peace in Darfur is not an

---

[52] Andrew S. Natsios, *Sudan, South Sudan, and Darfur. What Everyone Needs to Know* (Oxford: Oxford University Press, 2012).

objective that can be reached in the near future and writes: 'No successful peace process in recent history has included so many competing players or such exaggerated demands.'[53]

This broad description of the situation in Darfur and the rest of Sudan together with its representation on Google Earth can be understood as contributing to the *social gest* current at the times of our viewing of the images found in my case studies.

The major most recent source of easily accessible information about what is happening currently in Sudan, however, is the Satellite Sentinel Project Sudan, set up in December 2010 to monitor conflict as South Sudan prepared to hold a referendum in January 2011 on its planned secession from northern Sudan. Fighting has begun again between northern and South Sudan, with tens of thousands once again fleeing their homes in the northern sections of Southern Sudan to seek refuge in refugee camps.

Broken links, new links

The machine… is always at the junction of the finite and the infinite, at this point of negotiation between complexity and chaos.

Félix Guattari[54]

One aspect of the chaos of 'Crisis in Darfur' is that some of the links to video information have been broken or changed. It is easy to assume this is because contributors of these videos have repositioned them on their own sites and have not provided updates on the Google Earth site. However, some videos seem simply to have been removed, for example, the video of *janjaweed* confessions that took place 'in a tent' and which so far I now cannot find anywhere else on the WWW. This repositioning or deleting of links also applies to photographs and testimonies that originally were available via 'Crisis in Darfur'. As far as I can see at the time of writing, information on

---

[53] Natsios, *Sudan, South Sudan and Darfur*, 191.
[54] Félix Guattari, *Chaosmosis. An Ethico-aesthetic Paradigm*, trans. Paul Bains and Julian Pefanis (Sydney: Power Publications, 1995), 111.

the conflict ceased to be updated on the 'Crisis in Darfur' site by the end of 2009, although when a flame icon links back to an institutional site, for example, on USHMM, it is possible to use that site to find updates up until 2009 at least.

So one of the many interesting things about the site is that it seems to be perceived now by the site's creators to have to a large extent served its purpose. It is primarily an archive of the conflict in Sudan and neighbouring countries during the time period of 2007–2009. When I have recently gone to 'Crisis in Darfur' on Google Earth on my home computer (6 May 2014), the background to the layer is white without the topographical imagery from Google Earth that exists outside the layer. Compared with the image inserted earlier in Figure 4, it is clear that the site looks downgraded in its lack of positioning onto the kind of topographical information available everywhere else on Google Earth. It could be simply suffering from processes of updating: and this is worth considering as Google Earth processes images and texts in forty different languages.[55] Problems of low bandwidth could be another reason why images might appear degraded during some searches, although this would not explain why the area in Google Earth directly surrounding the site can be viewed with all the topographical detail usually associated with it, while the 'Crisis in Darfur' layer appears degraded. When viewed on my computer at my university office, however, on the same day (6 May 2014) the site appears with the topography intact. Bandwidth may well account for this difference, but it is interesting that such a difference can happen because this means that different times and computer hardware can create such different viewing experiences of an important information heavy site.

My overall assessment of the layer is that it has changed to quite a large extent over the last two years; it also seems to be continually altered, at least by some newer tagging icons that do not appear to be part of the original layer. For example, when I went into the site on 3 May 2014, there were information icons that linked to web pages in the Google Earth Community. Photographs and information that were linked previously to the USHMM site are now linked via these Google Earth Community pages.

---

[55] GEOE Interview, 14 February 2014.

A site of such significance can of course change due to technical issues, but it can also change over time according to what information stakeholders want to contribute to the site, and when. And this is what has happened recently with 'Crisis in Darfur'. When looking (6 May 2014) to replicate screen captures of web pages previously taken in 2010, I found links to the photographs I was researching, but the link took me back to a USHMM web page that said the Museum had recently changed the design of their website. So links that previously brought up photographs and showed them embedded in 'Crisis in Darfur' no longer existed there. The context was changed, the images were swept off into other locations on the WWW. The USHMM recommended going to their home page and searching there, but you would need to know what you were looking for. The 'stumble upon' factor does not therefore operate in this case.

A new kind of history

'Crisis in Darfur' is now positioned as a flagship site for how Google Earth Outreach promotes awareness and advocacy; it is now an archive of the conflict in a particular time and space.[56] This shift from active documentation to archival practice is still an important act of advocacy. It is a kind of documentation that provides us with a new kind of history – one that is created from content from many different sources that are nevertheless contained within a coherent form. What we call this form is not yet clear. It consists of what information we are given to see on screen and what we search for in the context of that information. In other words, this kind of history text comprises our searching a website. It is also crucial to remember that information we search for and respond to has been put on the WWW by a wide range of people drawing on an also wide range of sources. This information then can be the result of careful scholarship, journalism and institutional, corporate and individual research as well as for self-promotion. The promotion of an interpretation of actuality is a bias that is interrogated by scholars of historiography. Digital documents are always open to change

---

[56] GEOE Interview, 14 February 2014.

in ways that are often difficult to research except through forensic attention to sources and time dating. 'Crisis in Darfur' is a good illustration of the challenges that digital technologies now confront us with concerning formats of history, historical sources and archives.

It is fair to say that the significance of 'Crisis in Darfur' as a conduit of current information on genocide in Darfur has been overtaken by other major human rights activism websites that appear to have ceased their contributions of more current information to the Google Earth layer. Such sites include those owned by the USHMM (see Figure 5.1), Amnesty International, Doctors Without Borders, Human Rights Watch, Eyes on Darfur, Save Darfur and the newer site Satellite Sentinel Project, launched in 2010 and run initially from the Harvard Humanitarian Project until 2012. Google Earth assisted this latter site in its early days by contributing its map-making technology, thereby granting access to satellite imagery data from DigitalGlobe, the same data company used by Google Earth.

Information from 'Crisis in Darfur' has always been very difficult to access, even during its halcyon days, with the flame icons hiding many of the links to photographs, witness testimonies and videos. Nevertheless, the layer 'Crisis in Darfur' still survives, showing up quite clearly on Google Earth, still tracing a place and time on Earth of a savage war. The fact that it remains as an artefact of a particular period of time illustrates how past events can fade into static icons of reified memory on the WWW – where the illusion of continuing time can operate whilst its ongoing passage is nevertheless erased. Whilst 'Crisis in Darfur' itself offers a textual *gest* of obfuscation or loss, it can also be understood to contribute towards a wider and still current *social gest* that exists within a broader and continuing process of secrecy, grief, confusion and anxiety.

Celebrity activism and the Satellite Sentinel Project

Before going on to discuss my case studies, I need to note a phenomenon that makes a significant contribution to the context of both sets of images: celebrity activism. The entry of celebrity activists into the advocacy arena of the Darfur conflict and consequently into any discussions about the Google Earth site introduces another factor into the mix of embedded power relations and the kind of information and imagery available on these sites about distressed people in conflict zones. Virgil Hawkins asks the relevant question: 'Just how much of an impact does such celebrity activism have?'[57] He goes on to discuss and collate information on media representations of the Darfur conflict and how they correspond to various kinds of celebrity activism including that of actors George Clooney and Angelina Jolie.

The Darfur conflict began to draw world attention primarily in the United States. Media interest began in earnest in 2005, before significant celebrity activism. Torchin describes an example of how some of the groundswell of celebrity activism for the Sudan conflict emerged from a pre-release partnership between Amnesty International USA and United Artists in order to produce an 'educator's guide' to the film *Hotel Rwanda* (Terry George, 2004). This film tells the story of Paul Rusesabagina who saved the lives of over a thousand people during the 1994 Rwandan genocide, by protecting them within the hotel in which he worked, the Hotel Mille Collines.

Don Cheadle, the actor who played Paul in the film, used the attention given to this film and the genocide it portrayed, as well as the ongoing process of reconciliation in Rawanda, to draw attention also to the genocide happening in Sudan.[58]

Six years later, the SSP was created largely as the result of the concern of several celebrity activists about the Darfur conflict. The most prominent and active of these is George Clooney. This is a quote from the 'about' web page:

> On a trip to southern Sudan in October 2010, George Clooney and Enough Project Co-founder John Prendergast had an idea. What if we could watch the warlords? Monitor them just like the paparazzi spies on Clooney?
> http://satsentinel.org/our-story/george-clooney (18 January 2013)

This webpage also includes the following comment on Clooney's trip to South Sudan in March 2012:

[57] Virgil Hawkins, Virgil, 'Creating a Groundswell or Getting on the Bandwagon? Celebrities, Media and Distant Conflict', in *Transnational Celebrity Activism in Global Politics*, eds L. Tsaliki, C. A. Franonikolopoulos and A. Huliaras (Bristol: Intellect and Chicago University Press, 2011), 88.

[58] See Torchin, *Creating the Witness*, 168–169, and Notes 66 and 68, 242–243.

This was no media stunt – he traveled like a journalist, staying in tents, riding in the back of trucks, and meeting survivors, policy-makers, and militants along the way.

Clooney's activism/patronage also extends to the website 'Not on Our Watch', a journalistic website that was a precursor to the SSP:

> Drawing upon figures with uniquely powerful voices, we develop advocacy campaigns that bring global attention to international crises and give voice to their victims. We target mass media and international press, and engage world leadership.
>
> http://notonourwatchproject.org/what_we_do (18 January 2013)

'Not on Our Watch', one of the first major sites to address the Darfur genocide, was initiated and founded by Clooney and other Hollywood celebrities. It is linked to many other sites of course, including 'Look to the Stars: The World of Celebrity Giving' – representing the lower end of how online celebrity activism works and demonstrating the unfortunate kind of contextual fringe that is nevertheless part of the *ecos* of my two case studies.

Two sisters: Trauma-Kalma

Figure 5.1 'GIRL WITH BABY SISTER. Girl with traumatized baby sister. The baby has not made a sound since the day their parents were slaughtered and the village burned.' Photograph by Mia Farrow ©, screenshot on 13 December 2010, USHMM ©, Google Earth ©. Google and the Google logo are registered trademarks of Google Inc., used with permission.

This photograph was taken by Mia Farrow, a film actor who is also a photojournalist and Goodwill Ambassador for UNICEF. The photograph is also available on Farrow's own website amongst many more that she took during her several journeys into Darfur.[59] Her website http://www.miafarrow. org/ provides photographs and a poignant 9-minute video *The Darfur Archives*. As with Clooney, Farrow is a serious human rights activist who is also a 'Hollywood' celebrity. Farrow's site is primarily concerned with human rights activism but nevertheless borrows from her fame as an actor for its visibility on the web and contributes to her status as a frequently invited speaker on genocide. Farrow notes on her Flickr site that the photograph was taken on 13 June 2006.[60] She names the photograph 'Trauma-Kalma'. Kalma[61] is the refugee camp in which the photograph was taken. This exact location is not given on 'Crisis in Darfur'; closely examining the Google Earth layer, it is probably near Al Fashir, the capital of North Darfur.

A personal reading

Both children look directly into the camera lens. This gaze recalls Willemen's 'fourth look', his idea of a cinematic 'direct address' that demands an engagement with the viewer because the viewed show they know that they are being looked at. This gaze confronts the morality of my viewing, I am isolated from these people in both time and space: how can I make sense of the connection I feel with them unless I pay them the respect of my close attention?

I realize that I want to describe these two sisters as 'lost'. I don't know their names and their photograph is almost lost too in the obscure, crowded image space of 'Crisis in Darfur'. I do not know where exactly they are

---

[59] See http://www.miafarrow.org/images/galleries/darfur/index.htm (accessed 24 January 2013).

[60] See http://www.flickr.com/photos/30238868@N08/7116524777/in/set-72157629542370912 (accessed 24 January 2013).

[61] For more information on the Kalma Refugee Camp, see Jonathon Steel, 'Violence Flares in Darfur's Kalma Refugee Camp as a New Cycle of Persecution Begins', *The Guardian* (Saturday 27 October 2007): http://www.guardian.co.uk/world/2007/oct/27/sudan.international (accessed 24 January 2013).

physically and emotionally, both when the photograph was taken and when I am looking at them; they are lost as to what is going to happen to them; the photograph most definitely lost in the overcrowded community of other Web images (including tagged tourist photographs!) captured from within the continuing genocide in Darfur. Trauma-Kalma is not a simple unitary image because it lies within 'Crisis in Darfur's swarm of maps, testimonies, satellite images, all constantly linked and unlinked, and linked again somewhere else on the WWW.

Straight away, another connotation occurs: one that drags the children into the viewing domain of global celebrity. Farrow has in the past attracted quite a lot of publicity about her nurturing of children – adopting several as well as the ones she shares biologically with life partners. Consequently, there is a glimpse of an idea that these children are being cared for by her via her camera lens and her background as a caring mother/activist. Looking closely at the photograph, my attention focuses tightly onto the faces of children although they are captured via a mid-shot. The older child looks at us with what I perceive to be a quiet but defiant cynicism and suspicion. The baby lies on the older girl's back; the baby's gaze embodies distress, anguish. The stereotype of children's innocence, helplessness is ruptured by this image. The older child carries her sister; she is an active, not passive victim.

What does this image of two children's faces perform towards the viewer when she clicks on its web link? If the children are helpless, so are the adults around them and in the wider context, so also are the governments, aid agencies and world communities that watch and intervene in ways that seem to move too slowly and uselessly to achieve changes for individuals such as these children. They have been uprooted from their village and they have witnessed violence against their parents, their protectors. Nothing can change that.

As I continued to trawl through many but nevertheless only some of the other Web images that show the catastrophe of Darfur, I came to think of Farrow's photograph of the girls as a close-up of the genocide. It does not denote explicitly the horrors of dead and tortured bodies of this genocide; its caption and context is needed to locate where the girls are and why they are being photographed. Yet within the context of their belonging to a much wider community of images of the conflict in Darfur, I am able to bring a compassionate gaze and perhaps also a state of witnessing, to the wider parameters of the genocide as it is mediated by the WWW. This mediation occurs through a human rights website, the medium of photography and the distributive powers of the Web.

The photograph certainly provides an example of the subaltern status of traumatized children that Debbie James Smith addresses in her article: 'Big-eyed, wide-eyed, sad-eyed children: Constructing the humanitarian space in social justice documentaries'.[62] My understanding of this photograph also plays into another problematic of power relations. The politics of my focus on the girls' faces needs to be addressed further in order to describe more of the performative processes that inform the flows of power at work through this photograph. In the words of Deleuze and Guattari, '[t]he face is a surface: the face is a map'.[63] Children's faces usually are unlined, canvases for experiences to be written on as the children grow; the maps of these children's faces tell of a knowledge no one would want to share. But the beauty of the photograph itself and my curiosity nevertheless hold my gaze. What does my gaze mean for the power relations between the sisters and me?

John Erikson draws on Deleuze's and Guattari's critique of facialization for his discussion of performance practice. He notes that the face is never 'represented' as such but is an entity of 'becoming something else', a site of flux that resists definition:

> [Deleuze and Guattari] posit an idea of *experimentation* instead of interpretation – no repetition, no representation, no signification, but always a becoming something else. For them signification and subjectivation are two strategies of social and state power over bodies continually trying to evade definition.[64]

---

[62] Debbie James Smith, 'Big-Eyed, Wide-Eyed, Sad-Eyed Children: Constructing the Humanitarian Space in Social Justice Documentaries', *Studies in Documentary Film* 3, no. 2 (2009): 159–175. doi: 10.1 386/sdf.3.2.159/1.

[63] Gilles Deleuze and Félix Guattari, *A Thousand Plateaus. Capitalism and Schizophrenia*, trans. Brian Massumi (Minneapolis, MN: University of Minnesota Press, 1987), 170.

[64] John Erikson, 'The Face and the Possibility of an Ethics of Performance', *Journal of Dramatic Theory and Criticism* XIII, no. 2 (1999): 6–7.

Further relating these ideas to digital images of the face, Tim Lenoir says the following in the context of experimentation: '... facialisation involves a certain excess over the framed image, one that catalyses a properly bodily affective response.'[65] So through my response to the girls' faces I am extending my bodily awareness of the unknowable, evolving situation they are in – using Baer's term, an awareness of their image's 'ungovernability'.[66] Within this conceptual context, I can identify further the intimate 'community of images' in which this image of the two girls sits and the politically textual power that this image exerts over its content.

The embedding of photographs of trauma particularly within Google Earth and the SSP indicates how humanitarian image-making now strongly relies on surveillance images constructed via remote sensing satellites; and also how images of people can be lost in the maze of other images and information even though they all might be contained within a single site. As noted earlier, despite its affective power, however, and its evocation of compassion in the viewer, such an image as Trauma-Kalma can still act as an illustration, an icon of nameless child victims of a conflict far away from me. Questions of agency, for both viewer and viewed, can become or at least seem moot when confronting the enormity of this genocide and its legacies. This illustrative capacity is even more pronounced in my second case study 'Smallest Witnesses'; and so again it becomes clear that images demand careful and prolonged to achieve a moral perspective.

'Smallest Witnesses: The Crisis in Darfur through children's eyes': Friday, 3 June 2005

This case study consists of children's crayon drawings and the ways in which they are web-linked to both videos and photographs of the children holding the drawings and of the drawings themselves. They were drawn by refugee children from Darfur in early 2005. The drawings are of events

---

[65] Tim Lenoir, 'Affect as Interface: Confronting the "Digital Facial Image"', in *New Philosophy for New Media*, ed. Mark B.N. Hansen (Cambridge, MA: MIT Press, 2006), 133.

[66] Ulrich Baer (2002) in W.S. Hesford, 'Documenting Violations: Rhetorical Witnessing and the Spectacle of Distant Suffering 1', *Biography* 27, no. 1 (2004): 114.

Figure 5.2 Screenshot from the Human Rights Watch web page 'Failing Darfur' http://www.hrw.org/sites/default/files/features/darfur/smallwitnesses/drawing.html captured on 29 January 2013.

that happened to them, their families and neighbours during raids by the *janjaweed*. The children lived in camps near the Chad/Sudan border. Researchers into sexual violence during the Darfur conflict gave paper and crayons to the children mainly to keep the children occupied while they interviewed the adults. The following quote from the Human Rights Watch web page well describes what the children drew:

> Over the following weeks of the investigation, these violent scenes were repeated in hundreds of drawings given to Human Rights Watch, depicting the attacks by ground and by air. Children drew the Janjaweed over-running and burning their villages and Sudanese forces attacking with Antonovs, military helicopters, MiG planes and tanks. With great detail, children drew the artillery and guns they had seen used, including Kalashnikovs, machine guns, bombs, and rockets. They also drew the attacks as they had seen them in action: huts and villages burning, the shooting of men, women and children, and the rape of women and girls.
>
> http://www.hrw.org/sites/default/files/features/
> darfur/smallwitnesses/intro.html (29 January 2013)

The drawings were also photographed by Farrow; some of the photographs also show some children who drew them. These children with their drawings appear in the 9-minute video embedded via vimeo on Farrow's website: *The Darfur Archives,* http://vimeo.com/16116943 (29 January 2013).[67]

The screenshot captured in the Human Rights Watch website and shown in Figure 5.2 is from the most vivid account of the images and their making. This account is by the researcher who brought the crayons and paper into the camps. In a 5 51-minute video, Dr Annie Sparrow describes what the drawings are about. This is also the same video in 'Crisis in Darfur' that is made available via a link to the USHMM 'Speakers Series pages' http://www.ushmm.org/genocide/analysis/details.php?content=2005-06-03 (29 January 2013).

My personal reading of these images is as various and even as numerous as the drawings themselves. The number of them, the many different situations they represent, the number of children who drew them and the many places on the Web where we can view them – these factors all confound any clear apprehension of what they actually mean when viewed as photographs or through video or an annotated slide show. For me, the slide show actually provides the most immediately transparent account while the short video including Sparrow's witnessing of the images is the most urgent – with its documentary-style talking heads, response shots and image inclusions of the drawings themselves. This video is an interesting example of the embedding of the actual viewing of video within a website itself that was not available before HTML5:

> With HTML5... Video coded into the web page enables a dynamic relationship to static and live web data. In the same way that a hyperlink allows a connection between a word and another location on the Internet, so now such a connection can be made from a point within a video timeline or image.[68]

---

[67] Images from this series also appear directly via video on the following web pages belonging to the USHMM, also available on YouTube and Human Rights Watch websites, respectively.
http://www.ushmm.org/genocide/analysis/details.php?content=2005-06-03
http://www.youtube.com/watch?v=uMdyhFaxTKE
http://www.hrw.org/sites/default/files/features/darfur/smallwitnesses/drawing.html
(all accessed 25 January 2013).
[68] Dovey and Rose, 'We're Happy and We Know It, 164.

HTML5 video embedding does away with the distraction of moving to another viewing site, promoting a more focused style of immersion in the content of a video, and this is important for the kind of viewing needed for 'Smallest Witnesses'.

When I look at the drawings themselves I need to pay even more and closer attention to them than I do with Kalma-Trauma, even if my interaction with them is still focused within the screen space of a website page. This is because I need to translate for myself the stylistics of what the children drew, to interpret for myself what meaning the children might have been intending to express. To pay respect to these children, I need to go beyond the annotations and narratives that have been packed around them via video, photography and slide show. I need to own them as images that first and foremost affect me and demand action of me (even if this action is the affective emotion of compassion that leads to my writing academic articles about them). I can thereby avoid what Cubitt describes as the dangers of relying on annotated explanations and declarations of Web viewing exhaustion that justify 'a sentimental paroxysm of the black hole of meaning' that he claims is an assumed naïveté which acts 'to propose a mesmerised quietism in the face of the plethora of images [and] is to renounce responsibility for the future... acquiescence in the extinction of whole peoples'.[69]

These drawings need to be understood both as a series of by now iconic illustrations of genocide and also as individual statements of personal grief and bewilderment. Most of the drawings tend towards naturalism, and the children even attempt somehow to depict physical trauma using colour to denote people and events. Indeed colour is an interesting factor in my relating to the drawings. For example, in the drawing shown in Figure 5.2, the use of blue and finer line drawing in black connotes more attention to detail (and therefore implied accuracy?) rather than the bright colours used when one child draws an explosion. In the drawing shown in Figure 5.2, I am taken directly into the experience of the person drawn with her/his hands raised in a gesture of helplessness and horror. My understanding of this

---

[69] Cubitt, *Digital Aesthetics*, 46.

drawing and the others also depends on a confrontation with what I know of other children's drawings. As a parent I have a particular viewing context; these drawings resonate with my memories of more innocent drawings by children.

My viewing is a personal one of witnessing clearly authored texts, not one of the disembodied surveillance that can occur if images are not obviously embedded within the context of human rights websites. Each drawing itself constructs a performance of individualized *gest* of bewildered fear that is authored by scared, traumatized and probably angry children. The various and further embeddings of the drawings within the Web further perform a *social gest* still current in Sudan – of fractured innocence, of childhood transformed by violence.

In order to describe the intimate *ecos* of these drawings I actually need to include all of them that are available to me via the Web. Since I originally found them in 'Crisis in Darfur', in the context of my viewing history this site constitutes their intimate, domestic *ecos*. When I move away from this site to the other websites and pages that display the drawings I am moving towards another *ecos* that links the drawings to a broader discourse on conflicts in Sudan and even further on to websites of other conflicts. The jangled web space of YouTube and CitizenTube[70] then also constitutes the outer limits of their *ecos*: with all its videos and photographs that have some relevance to the meanings and histories of the drawings.

Ethics of mediation and a 'politics of dislocation'

Markers of the shift from older to newer media formats in documentary practice include then the paradoxical sense of both increased social isolation (in Turkle's sense) that arises from our private use of personal computers to find information about people far away, and also the sense and knowledge that our computers are in fact our machinic companions: that any knowledge we gain is in fact a result of a new kind of subjective positioning of ourselves

---

[70] For more examples of videos and photographs of the Darfur conflict in YouTube, Torchin, *Creating the Witness*, 196–215.

as viewers of digital texts. In his interrogation of the relationship between subjectivity and machines of enunciation, Félix Guattari asked the following question:

> how do we reinvent social practices that would give back to humanity – if it ever had it – a sense of responsibility, not only for its own survival, but equally for the future of all life on the planet…. love and compassion for others?[71]

In other words, how can I explain and experience my searching for human rights violations information on the WWW beyond my looking for what Gregory calls a 'sight-bite of global mayhem'?[72] I suggest that any address to websites that show people whom we do not know asks this ethical question of us the viewers, the witnesses, the participants. Any interaction with any website is never innocent or transparent, even more so with those that claim to show actuality.

In their book *Electronic Elsewheres*, Berry, Kim and Spigel ask: 'What is an "electronic elsewhere"?' They say that they use this term

> to emphasize the idea that the media do not just represent – accurately or inaccurately – a place that is already there. Rather… places are conjured up, experienced, and in that sense produced through media.[73]

The above quote is almost a truism by now: even the general public has become aware of how their representations of self can be manipulated and distorted on social media platforms. Belatedly for many, we now realize how the models and simulations of place and time are given to us in cyberspace 'as if' they exist as physical entities and yet might only exist in the actual (if virtual) worlds of our imagination and speculations – how fantasy and the actual, fiction and non-fiction are interwoven through each other in cyberspace in ways that contest any unravelling. However, it is still necessary to begin this quest to understand how the representational processes of the

---

[71] Guattari, *Chaosmosis. An Ethico-aesthetic Paradigm*, 120.

[72] Sam Gregory, 'Transnational Storytelling: Human Rights, WITNESS, and Video Advocacy', *American Anthropologist* 108, no. 1 (2006): 195–204.

[73] Chris Berry, Soyoung Kim, and Lynn Spigel, 'Introduction: Here, There, and Elsewhere', in *Electronic Elsewheres. Media Technology and the Experience of Social Space*, Public Worlds, eds Chris Berry, Soyoung Kim, and Lynn Spigel (Minneapolis, MN: University of Minnesota Press, 2010), vii.

WWW work. The idea of 'electronic elsewheres' is a useful one for considering how images on the web perform within the power relations of a global society. In terms introduced in this chapter, they perform a *gest* which contributes in turn towards wider understanding: a *social gest* that allows conclusions (albeit always shifting and often tenuous) to be drawn about 'the social relationships prevailing between people of a given period'.[74]

Referencing Larry Grossberg's idea of a 'politics of dislocation' (1999), David Morley says the following with regard to how our sense of time and place is changing via the 'elsewhere' of media:

> It is now commonplace that the networks of electronic communication in which we live are transforming our senses of locality and community – and in this context it has been argued that we need to develop a 'politics of dislocation' that is concerned with the new modalities of belonging that are emerging around us.[75]

His interest, and mine in this chapter, 'is what all this does to the relation between the media and the domestic sphere – conventionally the place of belonging, par excellence'.[76] When Morley's comments are applied to the digital domain, they open up a useful discussion on how our perceptions of web-based depictions of people can become disjointed from their origins in time and place – how they travel backwards and forwards in the private/ public domain of the WWW and the various other media that are distributed via the WWW. Since the photograph of the 'Girl with Baby Sister' and the 'Smallest Witnesses' drawings are embedded in 'Crisis in Darfur', they also are embedded within Google Earth, a platform invoking an infinite process of transference between human subjectivity and change. As such they are embedded in animated models that have been derived from satellite images obtained via remote sensing. This process and even its naming as 'sensing' implies affect and is part of the *ecos* of any images contained in Google Earth. By definition our senses are engaged even at such a distance; this affective status of satellite images from space further infers an illusion of indexical status

---

[74] Brecht, *Brecht on Theatre*, 139.
[75] David Morley, 'Domesticating Dislocation in a World of "New Technology"', in Berry and Spigel, *Electronic Elsewheres*, 3.
[76] David Morley, 'Domesticating Dislocation', 3.

to Google Earth and the images documented in 'Crisis in Darfur', reinforcing our interpretation of them as representations of actuality.[77]

The *social gest* of a new political aesthetic

Massumi's conceptualization of the 'aesthetico/political' in 'occurrent art' well encapsulates some of the problems in describing the various levels of *ecos* in which my case studies can be located. As he discusses the issue of containment of meaning in an artistic expressive event, he notes the following:

> The compositional problem cannot be addressed without at the same time addressing the problem of relational co-habitation, which is *ecological*: which extra-elements will be admitted into the symbiosis of compositional co-immanence? Which will be treated as predators or competitors and be held at bay? How, and at what proximity or distance, to what follow-on effect?[78]

These questions directly address the problem of linking the *ecos* of images to their actual content. I suggest that the viewing process of always moving back to the specific content of an image and then outwards to related content on the WWW constitutes a particular *social gest* that is relevant to the current moment of documentary culture. It is one of iconography, universalism, moral confusion and the possibility for new 'grand narratives' of human rights activism: lots of noise and spectacle obscuring kernels of effective and affective action. This *social gest* also reflects a belief that watching is to witness is to act, a shamelessness about activism and virtual tourism, an obfuscation of what is effective action. So much information sometimes so badly delivered that it can become misinformation.

---

[77] For further discussion on affect and remote sensing, see Caroline Bassett, 'Remote Sensing', in *Sensorium. Embodied Experience, Technology and Contemporary Art*, ed. Caroline A. Jones (Cambridge, MA: MIT Press, 2006); see also, Lisa Parks, *Cultures (Satellites and the Televisual) in Orbit* (Durham: Duke University Press, 2005); and Catherine Summerhayes, 'Embodied Space in Google Earth: *CRISIS IN DARFUR*', *Media Tropes* 3, no. 1 (2011): 113–134.

[78] Massumi, *Semblance and Event*, 155.

Markers of the shift from older to newer media formats in documentary (and journalistic) practice include then the paradoxical sense of both increased physical isolation arising from our private use of personal computers to find information about people far away, and also the feeling and knowledge that our computers are in fact our machinic companions: that any knowledge we gain is in fact a result of a new kind of subjectivity derived from this companionship. As quoted earlier, in his interrogation of what he names a new 'ethico-aesthetic' in the context of the relationship between subjectivity and machines of enunciation, Guattari asked 'How do we reinvent social practices that would give back to humanity – if it ever had it – a sense of responsibility?'[79] I suggest that any address to websites that show people whom we do not know asks this question of us the viewers, the witnesses, the participants. Any interaction with a website is not innocent or transparent, and currently we can only answer this question in a speculative fashion.

Our reliance on web imagery sourced from satellites, cameras in the 'outer space' of our home galaxy, epitomizes our dilemma of simultaneously experiencing the differences between two spatial markers: that between the near and the far and that between environments of horrific distress and the comfort of our own homes. Then of course there is the time gap. Although digital communications can verge towards the instant, images embedded in websites usually come from a time past, however short that time might be. That people, situations and places change through the vector of time is also a truism but one that is useful for reminding us how such websites and images also embed representations of people in times past. Problems of victimology and spectacle also impinge on our awareness of how these images work on our senses, our imaginations and on the broader political sphere. As Alexander Galloway asks:

Would photographs of suffering move us? And if we are not moved, are we to blame?[80]

---

[79] Guattari, *Chaosmosis*, 120.
[80] Galloway, *The Interface Effect*, 89.

Michael Renov poses a similar question: 'What kind of responsibility do you bear for that other?'[81] These questions also inevitably arise when we come to the Other via the WWW and the fascination we have with images of other people's lives far away in time, place and social situation. The *ecos* of these images is complicated, often obscure and finally ineffable. Its existence needs to be marked, however, as one way of respectfully acknowledging the makers of the images, their documentary content and the people implicated in the viewing of this content. To 'fold' something is a gentle action and requires respect for the fabric that we are folding; to fold our findings and contemplation of people and places found in Google Earth demands a gentle yet rigorous exercise of our ethically driven ability for compassionate, active witnessing.

---

[81] Michael Renov in 'Collaborations and Technologies. Stella Bruzzi (interlocutor), Gideon Koppel, Jane and Louise Wilson in Conversation', in *Truth or Dare. Art and Documentary*, eds G. Pearce and C. McLaughlin (Bristol: Intellect, 2007), 77

6

In Conclusion: Humanist Trajectories in Google Earth

The machine is not an it to be animated, worshipped, dominated. The machine is us, our processes, an aspect of our embodiment.

Donna J. Haraway[1]

In this concluding chapter, I develop further the idea that Google Earth can be understood, comprehended to be 'a cultural performance'. I do this to give an overview of how I understand Google Earth as a global player in society's quest for knowledge about humanity. This behavioural trope of cultural performance was introduced in Chapter 3 as specifically referring to one stage in the anthropologist Victor Turner's concept of 'social drama'. Here I consider how this idea of a social drama can be used to think about the social context of Google Earth. I ask the question: can we analyse Google Earth as playing a crucial and critical role in the social drama of humanity's now globally based awareness that the degrees of separation are dwindling exponentially between humans, other life forms and the Earth itself. If we consider then the whole phenomenon of Google Earth as a text that performs necessary cultural functions, then we can use the language of performance studies to note its potential for social critique.

Our re-positioning of the status of humanity as only one vulnerable species on earth has its beginnings in projects of the enlightenment. Now, in the early decades of the twenty-first century, the rise of the World Wide

[1] Donna J. Haraway, 'A Cyborg Manifesto: Science Technology and Socialist Feminism in the Late 20th Century', in *The International Handbook of Virtual Learning Environments*, eds J. Weiss et al. (Dordrecht: Springer, 2006), 146.

Web (WWW) as a kind of global language matrix has catapulted our understanding of how humans must be considered as a much more inclusive set of creatures. The WWW and the Internet have forced upon us awareness that we must re-calibrate our definitions of time, space and distance, if we are to be able to think clearly about environmental and conflict situations that confront us now. Indeed, we source our awareness of these situations more and more frequently via our use of the WWW and the Internet.

As a way to sum up Google Earth's overall cultural performance, I look here at how Google Earth performs a significant role in manufacturing humanity's global awareness of the way the world is for us in this present era. This specific global awareness depends on the ways in which we use Google Earth: in the contexts of the trivial, banal and sometimes malign, as well as in the profound, educational and socially aware ways we use Google Earth – all of which can occur within one browsing session by one single user. So Google Earth and those of us who create and use the platform and its images have associated burdens of responsibility to the world it represents. This world includes the people, creatures and places we view in its content, the software program itself with all its related technologies and artefacts and the screen vision that we are inventing as a result. For this concluding discussion about how we and our machines together create new horizons in perception, I draw again on ideas from Félix Guattari and Donna Haraway and introduce the term 'bioconvergence'. Its introduction might be belated but I have left bioconvergence till now because this term and what it means can act as a summing up of the momentum that drives us to somehow account for all the various, related, yet different ways we experience the world in sync with our personal digital devices.

Content–technology–screen vision

These words refer to how we imagine ourselves in the context of the cybernetic technologies that we now interact with as part of our daily lives. I have used the order I have placed these words in at the top of this section deliberately. In the act of using our computer screens we usually feel that the whole process begins with us. After switching the machine on, we look at the screen and we see what we want it to show us; we have control over the process in this way. If we change our ideas about how to order the experience, and put content first, technology second and screen vision third, the realms of affect, surprise, spontaneity and reflection become the most significant aspects of our experience. This is not in any way to say that the technology acts as a transparent filter to information about the world. In fact it is the very opposite of this. Content (representations, reflections, other people's comments) is constructed into digital texts via the technology. The world that content mimics or represents is there regardless of whether or not it comes to us as content of a digital text. So I am saying here that the medium is NOT the message but it is certainly the massage (a McLuhan thematic noted earlier in Chapter 2). We rely on all kinds of media to access information and other kinds of data content. And content will vary with different media; no matter how closely an adaptation might work, the same set of questions will attract differing although nevertheless related answers when located in literature, cinema, computer games, visual art, live theatre, dance or music.

In an obvious way, we are always looking for content and allow ourselves the luxury of treating our personal computers as direct conduits to the 'best', 'the highest quality' or at least the most useful information available. But if we look at this information/data as content and then adjust our knowledge about the world to fit this data, then it is also obvious that we are also open to surprises. Why? Surprises happen when the world (people, things, environment, other creatures) doesn't act the way we expect it to. The WWW is perhaps one of the most vibrant textual machines ever created because it can present content to us in so many different ways, borrowing from all the media I mentioned before. It offers surprises in the ways in which it shows us connections between different texts, different arrangements of content. The format of the WWW, its content makers and users, do this in such ways as to provoke new media researchers to claim that we must now consider our very perceptual processes to have changed.

Our ideas of space and time have changed, says Farman and many others cited in earlier chapters of this book. We now think about ourselves differently than before personal computers. Although cognitive approaches might emphasize screen vision as a new perceptive sense that has developed over time with a huge acceleration since the early days of cinema, it is the content itself, its design and delivery that I understand to be the crucial development offered by the WWW. We can now hear and see other people, connect to them, communicate with them, react to them at our own volition in our own present time and space, unrestricted by the differences between our and their historical or present time and space. In Chris Russill's words: 'Our ways of knowing and shaping environments, hitherto limited to local, national, or regional territories, would now apply to the "global scale." '[2]

Changing time in space

Google Earth is a vehicle for communicating annotated information about what the earth (as the world) looks like and what happens on planet Earth. This information is not coherent in form or style and has the same problematic status of a 'truth-saying platform' that necessarily uses technologies of representation, as do all other visual and audiovisual texts of non-fiction. Added to this is the paradox of the index, inherent to all non-fictional texts: Google Earth still has the utopian aura of transparent representation that cinema and photography possessed for at least the first 100 years of their invention – this aura is reinforced by Google Earth's iconic, brand image of our planet as a luminous spinning globe in dark space: Gaia in blue and white, spinning in space for our appreciation of its beauty. Add to this beauty the incredulity associated with our everyday sense of the ephemeral images taken in what is still for most people 'outer space' as well as seductive offerings to zoom and fly through and over our planet as never before in history or in any other virtual space – then it becomes clearer why Google Earth can be imagined and used as itself a creature of 'outer space': not quite a game, not quite an encyclopaedia, a tool for surveillance and control and correspondingly, a site of resistance to state and corporate surveillance. It is

---

[2]   Russill, 'Earth Observing Media', 278.

also for untrammelled imaginings of place and time, depending on the user's skill and knowledge of the software applications.

My investigation of 'Crisis in Darfur' interrogates in particular our awareness of this engagement in terms of 'distance' and its disappearance. I examine our active, interactive digital gaze for the perceptual framing of distance involved in the experience of Google Earth. Although others have quite rightly critiqued Google Earth for its military prehistory and for the 'fly over' aesthetic of its moving images, other responses are possible, and they raise new questions. In Parks' words regarding the televisual image: 'We need to devise ways of seeing and knowing difference across distances that complicate rather than reinforce militaristic and scientific rational paradigms.'[3] Can Google Earth, for example, also be considered a site for dialogic communication between us, the users of Google Earth, and the people we see within its content? Such a thought seems both fanciful yet obvious at first glance, until the enormity of the suffering that we see through the images of a site such as 'Crisis in Darfur' confronts us with a need to understand how and why we look at these images.

If our engagement with Google Earth is (inter)active, not passive and if, through our searching of this website we can actually construct the very images that we see then, as I have argued in Chapters 4 and 5, we can also understand ourselves to be 'performing' the content of Google Earth. It is also necessary though to remember that in using this globally available software program, we use language, both spoken and written text. Elsewhere I have argued (Chapter 4) that the vision we speak about in engaging with Google Earth is one that is pan-sensual, a version of proprioception that is associated with the particular kind of looking that we do with our personal computers, our 'screen vision'. And yet there is another way into the content we see on Google Earth: the language of thought that we see printed as text and sometimes even hear via linked audiovisual texts. We need also then to remember that the people who maintain and continue to construct Google Earth must wrangle at least forty different spoken languages.[4] In this sense, Google Earth is a

---

[3] Lisa Parks, *Cultures (Satellites and the Televisual) in Orbit* (Durham: Duke University Press, 2005), 107.
[4] GEOE Interview, 14 February 2014.

global phenomenon, yes, but one that is not achieved without a large degree of difficulty that is easy to discount when we are enclosed in our safe bubbles of common language groups.

Bioconvergence and our machinic companions

Another way in which to describe our embodiment in the event that is Google Earth, is to interrogate the term 'bioconvergence'. My use of this term lies in the context of thinking about human embodiment in digital space. In this sense, bioconvergence manifests through the way we perceive the world around us via digital technologies and how we assimilate these perceptions as 'everyday', 'normal' and 'natural' to our human state of being. The living human body assimilates as part of itself the powers of perception that some digital technologies seem to make possible. The bioconvergent perceptual effects that machinic technologies have on our human lives accumulate with every current development in these technologies.

While we explicitly incorporate digital machines into the very fabric of our bodies via medical prostheses and computerized procedures, we also use personal computer technologies as part of our everyday awareness of the world around us, of what is or seems to be now socially necessary for living both our private and public lives. Our computers enable and augment all earlier modes of representation. Kindles, tablets, smart phones, Blackberries and all other portable digital devices allow us a sense of being able to contain our lives, what is important to us, within sleek, aesthetically minimalist objects of plastic and microchips. I argue that through our use of personalized digital computers, at the same time we also are experiencing another much 'messier', emotive life aesthetic.

The complex communication patterns enabled by Web 2.0 software together with the shiny digital devices most of us carry around with us on a daily basis come together to allow contact between people in a new way. We experience this new kind of contact as immediate in both time and space. As noted throughout this book, with our new personalized digital intelligence, we perceive the far as near. There is now a growing intensity of bioconvergence between the human body's sensual, perceptual apparatus for thinking and feeling, and our machinic companions. My premise is that such a bioconvergent crossing of boundaries between far and near in the dimensions of time and space can also extend across the even more inaccessible boundaries of cultural difference. Through a bioconvergence between the personal computer loaded with Google Earth and our human perceptual apparatus, we can use the illusion of the ever-present to engage with people who are far away in distance, time and culture, in a more powerful way from what we have before.

Inside the experience of 'Crisis in Darfur', the term 'bioconvergence' is an apt description of such an interaction between my own body and the body of someone far away. Indeed, two ways of understanding how we can use the term 'bioconvergence' are evident in this scenario. There is the convergence between two biological bodies that are not inhabiting the same time and space. This is a convergence that has puzzled us since the advent of photography. There is also the bioconvergence between human and machine, whereby our vision of another person is not simply a machine-mediated gaze, but an emotionally sensed perception enabled by computer technology. This bioconvergence manifests as a corporeal presence that is not limited by existence in the same time and space. Can we speak then, of a bioconvergent vision which is not necessarily the emotionally detached gaze of the empowered at the dis-empowered?

The living human body assimilates as part of itself the powers of perception that some digital technologies seem to make possible. The bioconvergent perceptual effects that machinic technologies have on our human lives accumulate with every current development in these technologies. These perceptual effects include the joyous Dionysian illusion ascribed to by Kingsbury and Jones III when they describe Google Earth as 'the projection of an uncertain orb spangled with vertiginous paranoia, frenzied navigation, jubilatory dissolution, and intoxicating giddiness'.[5] Throughout this book, I have worked to attenuate the frenzy of this kind of utopian illusion by acknowledging its constraints and limits. However, it

---

[5] Kingsbury and Jones III, 'Walter Benjamin's Dionysian Adventures on Google Earth'.

is easy, and to a degree also helpful to enter into the poetry, the rhetoric of a software platform such as Google Earth, because its veil of transparency and sometimes sheer beauty also capture for us some of the ways it provokes us to use our imaginations. And the rigour of poetry is as strenuous as any other approach to knowledge when words prove difficult tools for describing the beauty of the prosaic world. The aesthetic of Google Earth's virtual globe and the mundane uses we make of it, the almost unbelievable breadth of potential knowledge that this software platform conjures – all these elements together form a new epistemological challenge.

Machines of sunshine

In Virilio's words: 'Images have become a new form of light.'[6] In a literal sense, when we gaze at our computer screens we experience the actual blue light transmitted to us from those screens, as we use our tablets for bedtime reading and our phones as light boxes on which we text the universe whilst sitting in a darkened theatre (sometimes lecture theatre of some kind!). Yet the idea of images as a new kind of light also offers some other kinds of insight into the affective drifts of awareness that our computers allow us to inhabit as we cruise the world via digital maps. Haraway describes the miniaturized computer machines on which we view these images as 'made of sunshine':

> Our best machines are made of sunshine; they are all light and clean because they are nothing but signals, electromagnetic waves, a section of a spectrum, and these machines are eminently portable, mobile...[7]

Haraway goes on to ominously contextualize her description of the ethereal nature of computers in the pain of underpaid workers who make them, and the uses that the military make of computer devices in missiles. Her juxtaposition of sunshine with scud missiles clearly communicates the danger in ascribing either utopian or dystopian interpretations onto the phenomenology of our

---

[6]   Paul Virilio (1988) in Anne Friedberg, *The Virtual Window. From Alberti to Microsoft* (Cambridge, MA: MIT Press, 2006), 151.

[7]   Haraway, 'A Cyborg Manifesto' 2006, 121.

machinic companions. As Haraway makes clear in the quote at the beginning of this chapter, we are responsible for them, not the other way around. What becomes important is that we are aware of the nature of our interactions with machines – how we act online and offline with machines and what they show us. Concepts of 'aura' and 'voice' are useful for exploring some of the ways in which we use personal computers to 'give voice' to ourselves and others, and for identifying the historical contexts of our screen vision. Art objects and artistic practice have always been significant markers for how we develop literacy in new technologies.

In a different voice

When comparing how society receives and makes use of digital art to the reception and social status of non-digital art, media theorist Michael Betancourt expands Benjamin's idea of the 'aura of a work of art'[8] and contradicts Benjamin's premise that this aura has to do with the historical material and process that goes into making an *original* work of art.[9] He sets himself a task to take the idea of aura of a work of art and apply it to digital art, that is to say, to digital objects. His discussion forms an explicit critique of the capitalist burdens of processes of production and consumption in the digital era. In the context of Google Earth, I think that one particular insight is of special interest. He understands a 'digital aura' that elides production with consumption, thereby negating the significance of the capital that lies behind the technology of making digital objects. In Betancourt's words,

> The aura of the digital separates the results from its technological foundation – the illusion of value created without expenditure...[10]

I agree that digital objects like Google Earth inherently are illusions of light and algorithm, playing into the various hegemonies that exist at the sites of

---

[8]   See Walter Benjamin, 'Work of Art in the Age of Mechanical Reproduction' (1935), in *Illuminations*, ed. Hannah Arendt, trans. Harry Zorn (New York, NY: Shocken Books, 1969).

[9]   Michael Betancourt, 'The Aura of the Digital', in *Critical Digital Studies. A Reader*, Second Edition, eds Arthur Kroker and Marilouise Kroker (Toronto: University of Toronto Press, 2013), 444–445 (433–436).

[10] Betancourt, 'The Aura of the Digital', 445.

their production and reception. And yet, as I have argued throughout this book, if we are to work towards understanding Google Earth's status in global society, we must also look for lines of possible resistance and subversion of those hegemonies of power. To reiterate, we do not just look at Google Earth: we position ourselves in space and time, we talk with each other and 'hear' each other via text, sound and we can communicate with a generosity of voice, using any media available in Google Earth.

I suggest that we can now say that we, who use personal computers, use them as voices to the world in the sense that we use our interactions with these intimate machinic companions to articulate what we think and choose to think. Our performances in the world via our complex modes of screen vision do not end with the screen. They become part of our decision-making apparatus; they are not simply conduits of information. As I have argued throughout this book, we experience the information they give us access to, as significant parts of acting out our own life experience with all its consequences. Yes, they enable us to connect with people, and no, this connection does not exist only as part of a network of information shared or denied. In using Google Earth we are embodied – humanly embodied – members in a community of vast proportions. Part of our knowing membership in this community relies on the technologies that give us access to it. Our machinic companions are so close and necessary to us in this venture that we can speak of these companions as also belonging to our embodiment within the community of people and places connected with us via Google Earth.

So what can we achieve through listening, watching, reading the Other via the maps and overlays of Google Earth? In 1982, feminist psychologist Carol Gilligan's seminal book, *In a Different Voice. Psychological Theory and Women's Development*, introduced the concept of 'ethics of care' into feminist theory, and so developed an idea of a gendered, relational 'voice'. She grounded the idea in the physical act of speaking, saying that 'speaking depends on listening and being heard; it is an intensely relational act'.[11] Gilligan conducted fieldwork interviews with men and women, and came to distinguish between

---

[11] Carol Gilligan, *In a Different Voice. Psychological Theory and Women's Development* (1982) (Cambridge, MA: Harvard University Press (1993), xvi.

the different kinds of information (Western) society gives to male and female children about how they should consider themselves as individuals and in relation to others. More recently, she has spoken about the speaking position from which we use 'a "different voice" – a voice that joined self with relationship and reason with emotion'.

For the purposes of this book – an investigation into Google Earth – Gilligan's insights are important because she goes on to define an 'ethics of care' (in contrast to an ethics of rights) that can also be used to describe the act of responding to another person who needs our care. She also describes this ethical position as 'the ethic of responsibility [which] rests on an understanding that gives rise to compassion and care'.[12] Gilligan gives us here another point of access for working out how we can describe the compassionate response we have to people we see via Google Earth, and for describing how this compassionate response might be useful to the public good.

Our acts of observing, caring and responding in some way are of themselves valuable as social acts that ethically connect us to others. This is not to say that a useful response is simply thinking 'dear dear isn't that awful (and I'm glad it isn't happening to me)'. Useful responses to visions of trauma and catastrophe can be many and varied, with different impacts. I am saying here that in the context of an ethics of care, a significant and compassionate personal response is an action that also holds the status of public action. They become part of a public position, a confirmation that it is appropriate to feel empathy, compassion towards people whose presence is given to us via digital media. To understand its relevance to my argument about digital media, it is worth following the concept of ethics of care a little further.

Political scientist and philosopher Joan Tronto developed and expanded Gilligan's ideas of voice and care. In 1993, she published her book *Moral Boundaries* and included a chapter called 'An Ethic of Care'.[13] In this chapter she grounds her definition firmly within practice and in the power relations and morality within which the giving of care has been embedded. Tronto identifies four elements of an ethics of care and says that these give rise to

---

[12] Gilligan, *In a Different Voice*, 165.
[13] Joan Tronto, *Moral Boundaries: A Political Argument for an Ethic of Care* (New York, NY: Routledge, 1993).

'four ethical elements of care: attentiveness, responsibility, competence, and responsiveness'.[14] I suggest that all four of these are inherent to the affective emotional intelligence derived from compassion. The degrees to which we bring them into play via our own personal responses vary according to the force and emphasis we are able to give to them.

Love at a distance?

As noted before, artists often work at the frontier of new technologies. Although they might experiment without the rigour of empirical science, they closely engage with the nevertheless rigorous practice of finding out what technologies can do to and for human society, other creatures and the earth itself. And so we find the artist Susan Kozel's words, vividly expressing the melding of new and older forms of responsivity and responsibility that might evoke compassion via sites like that of Google Earth:

> Spontaneous compassion is not derived from axioms or rules; it arises from the demands of responsivity to the particularity and immediacy of lived situations. The virtual self, as decentered and spontaneous, performs and improvises within an underdetermined space. This sense of groundlessness wherein responsivity to the elements of a new system escapes habit and fosters new movement and ideas ...[15]

So, love at a distance? Is this possible in the sense that love is a compassionate engagement with the Other? In this book I have argued that we can imagine such a kind of engagement across time and space in opposition to previous *ennui* from other such repetitive imaging of people, wars and places that it seems we cannot affect, that exist so far beyond our own actual experiences. We will always be able to simply file the information gained from these images as documentary, truth-saying information – as documents to be filed so we know about them but do not have to be affected by them if we choose not to be.

---

[14] Joan Tronto, 'An Ethic of Care' (1993) reprinted, in *Feminist Theory: A Philosophical Anthology*, eds Anne E. Cudd and Robin O. Andreasen (Malden, MA: Blackwell Publishing, 2005), 252.

[15] Susan Kozel, *Closer, Performance, Technology, Phenomenology* (Cambridge, MA: MIT Press, 2007), 82–83.

Post-2005 and the emergence of the culture-defining sites of the social web, however, we have access now to a kind of compassionate engagement with people and places via digital imaging and imagining. We lay the foundations of this kind of experiences in our use of the web to communicate our intimate thoughts in Facebook perhaps, and in using Google Maps and Google Earth to imagine our own special, spatial fantasies and actualities of the world we live in. In this sense then, Google Earth not only offers a touch from the world beyond our physical self, it can also, in ways different to before, make it possible for us to own the stories of people and places in this world that exist beyond our embodied selves. In 2005, Lisa Parks commented on the very new rendition of satellite images into the realm of popular imaginations:

> Perhaps we could imagine the satellite as generating a kind of 'orbital pull', a metaphorical dislocation, a figurative removal from the zones of security and comfort in the world, forcing us to recognize the partiality of vision and knowledge and to embrace the unknown.[16]

Parks (2010) also notes that artists have led the way into how such popular imaginings and satellite imagery are currently being articulated by installation artists in both words and work. An example is Bassett's wonderful question about remote sensing offering the possibility of 'love at a distance'.

Artistic practice of sousveillance is subverting the surveillance/violence nexus of military practice towards a new one based on knowledge gained from interactive, haptic experience gained from a database of images from space. The interactive experience itself is not then merely to be thought of as simply accessed through satellite-sponsored, remote sensing technology. These technologies are a performative aspect of this experience; we engage with them as we interact, even if we do this via the agendas of corporations such as Google.com.

As another writer who interrogates Benjamin's idea of the aura of a work of art and applies it anew, Stahl claims that because of its history and the use of 'surveillance from the sky' technologies 'Google Earth has been unable to

---

[16] Parks, *Cultures in Orbit*, 91.

shed its martial aura'.[17] This may be so, but it is not the only aura available to this site, as illustrated by its various civilian uses as personal mundane records of wanderings over the earth and as a site for the re-presentation of human rights and environmental activism. The haptic nature of the knowledge gained from a site such as 'Crisis in Darfur' and the interactive combination of near and far vision given to us via Google Earth set up the conditions for compassion. As we search and find by happenstance or intention the terrible consequences of war, perhaps we are confronted with a need to think and imagine at a new level where the virtual is also recognized and known as embodied actuality.

In Chapters 4 and 5, I argued that 'Crisis in Darfur' is a moving, plastic, shape-changing subtext that we can use in many ways. Consequently, it can be understood as a particular kind of object: a medium for communication and representation. The human, the machine, compassion and grief – all of these are part of the experience of engaging with suffering people in Sudan. Political implications? Butler's words bring attention to the power of grieving in public, and how grief expressed communally can threaten hegemonies:

> Open grieving is bound up with outrage, and outrage in the face of injustice or indeed of unbearable loss has enormous political potential. It is, after all, one of the reasons Plato wanted to ban the poets from the Republic.[18]

Chapters 4 and 5 present Google Earth as a site that can elicit a responsive performance of compassion and that therefore Google Earth offers a certain kind of embodied, haptic experience that plays across our senses, which can be described as one that is socially useful, and to a degree necessary for further political action. Compassion is an act of knowing through recognition that the pain of another is a moral responsibility for the whole human condition. 'If people know, they will act accordingly.'[19] These simple words from Torchin are optimistic, but events do not always play out according to this optimism. Nevertheless, the question of compassion requires a consideration of

---

[17] Stahl, 'Becoming Bombs', 1.
[18] Butler, *Frames of War*, 39.
[19] Torchin, *Creating the Witness*, 1.

knowledge and power at an individual level, and of compassion as an active state that Marmor calls 'knowledge [which] is based on embodied subjectivity and that this form of knowledge is action'.[20]

Google Earth is surely a machinic vehicle for communications in the realms of both the sublime and the ridiculous – a vehicle dangerous to some agendas – that allows humans both to play and to grieve. The simulations in Google Earth augment our imaginations and perceptions of the world in new and powerful ways. Yet both the illusions and actual indexicality integral to remote sensing, mapping, animation and web technologies are contained within the time/space of a single digital object. Google Earth is now part of our perceptive apparatus, a new part of the ever-changing embodiment of the bioconvergent, human creature.

My second case study in Chapter 5, 'Smallest Witnesses', offers a useful example of how bioconvergence between personal computer and a human can result in an amalgam of bodily experience that cannot only be defined by the distance of 'far away'. The slide show and voice-over that is linked to the 'Smallest Witnesses' site in 'Crisis in Darfur' from the website of the United States Holocaust Memorial Museum (USHMM) is for me the most moving, affecting account about the drawings children made in their refugee camps – drawings depicting their parents and families being massacred. As noted in Chapter 4, when I see these children, they are in my live embodied space, my domestic space, and I have to deal with that.

In Chapter 5, I confronted the question of how the knowledge that I gain from such human rights media through my engagement with Google Earth differs from that accessed through other websites. How might this knowledge be different from that gained directly from activist websites, like that of the USHMM? I instigated the search for something I did not know was there and then followed through. I looked at this site with the same body that could drive the coast road on and off screen. I played Google Earth with my own body in order to understand places and situations I knew of previously either at a great distance (with little affect) or at too close a distance (with a great degree of affect). So I asked myself, how could I deal with an affective knowledge of

---

[20] Kathy Marmor, 'Bird Watching: An Introduction to Amateur Satellite Spotting', *Leonardo* 41, no. 4 (2008): 322.

people with whom I had no live, 'face-to-face' engagement? How to name such knowledge and how far could I go in defining this interaction as a new kind of face-to-face engagement mediated by the disappearance of distance, both actual and perceptual?

Here, the term 'bioconvergence' is an apt description for such an interaction between my own body and the body of someone far away. Indeed, two ways of understanding how we can use the term 'bioconvergence' are evident in this scenario. There is the convergence between two biological bodies that are not inhabiting the same time and space. This is a convergence that has puzzled us since the advent of photography. There is also the bioconvergence between human and machine, whereby our vision of another person is not simply a machine-mediated gaze, but an emotionally sensed perception enabled by computer technology. This bioconvergence manifests as a corporeal presence that is not limited by existence in the same time and space. Can we speak, then, of a bioconvergent vision which is not necessarily the emotionally detached gaze of the empowered at the dis-empowered?

In this last chapter, I have presented what might be seen as quite a utopian view of Google Earth, as perhaps some readers will say I have done throughout this book. This is not because I think that Google Earth is not open to use against the personal and public good, but because the possibilities for knowledge have not yet been exhausted. An ethics of Google Earth also needs to acknowledge that projects such as Google Earth offer great potential for even closer and more useful connections between people and places than are happening now. And this closeness, this being 'alone together' (in Turkle's terms) itself offers opportunities for empathic knowledge, for emotional intelligence as well as data collection.

A cultural performance

These words from Donna Haraway (first quoted in Chapter 1) well describe the interdependent relationship between our human selves and what we call our *personal* computers, large and small:

The 'eyes' made available in modern technological sciences shatter any idea of passive vision; these prosthetic devices show us that all eyes, including our own organic ones, are active perceptual systems, building in translations and specific *ways* of seeing, that is ways of life ....[21]

Haraway's naming of eyes that are 'prosthetic devices' akin to our own in that they are 'active perceptual systems' aligns well with Guattari's suggestion of a computerized 'machinic' production of subjectivity – another recent contribution towards how we can understand the ways that digital technologies work with us in the production of new modes of understanding and agency. Guattari asks the following:

What processes unfold in a consciousness affected by the shock of the unexpected? How can a mode of thought, a capacity to apprehend, be modified when the surrounding world itself is in the throes of change?[22]

Guattari speaks of subjectification in the context of psychotherapy, of the Conscious and the Unconscious; he confronts what he names as 'the massive development of machinic productions of subjectivity'.[23] My interest here is in Google Earth as a portal into a new form of comprehension that is not necessarily constrained by the anxieties and stress of creating a subject of self, but which comes about via a particular kind of bioconvergence – a convergence between our machinic companions (the personalized computer, satellite technologies, software applications) and our knowing bodies.

Google Earth presents us with many questions, a lot of information, ways of interpreting and adding to that information, possibilities for use to the good and for harm to humanity and Earth's environment. It is useful I think to use the trope of Turner's social drama as a way to finally describe here my very 'un-final' conclusions. We can think about Google Earth as a digital project that plays a specific role in our present digital era (early twenty-first century). The social drama is simply that we are now experiencing so much that is different in ways that are nevertheless eerily familiar. We still hear, touch, see and think as we always have, but added to these familiar senses we are now

---

[21] Haraway, *Simians, Cyborgs and Women*, 190.
[22] Guattari, *Chaosmosis*, 11–12.
[23] Guattari, *Chaosmosis*, 2.

experiencing each other, our environment and ourselves in spaces and times that have changed, and in ways that we have had very little time to prepare for. We are living in a time when, using Arthur and Marilouise Kroker's words, 'clear distinctions among flesh, machine, and images become increasingly difficult to ascertain'.[24] This lack of distinction between flesh, machine and image is echoed by and also related to our current difficulties in discerning the 'truth status' of digital objects on the WWW.

Google Earth is a project that certainly exists as itself, but it is also an avenue for thinking differently, for confronting what is different about how we relate to each other and the environment in the current era of digital technology. The connections that this technology offers to us are many and myriad, even if not necessarily unique in the digital realm. Despite our mapping of data connections, the sheer and seemingly ever-increasing volume of connections and the appetite we have for them point to unexpected nodes of information not dreamt of before. An example of this is my journey from Australia to Africa in Google Earth. Unwittingly I had ticked the box that allowed Google Earth to show the overlay 'Crisis in Darfur'. This coincidence opened up for me a new way of connecting with people far away from me in distance, time and situation. Cubitt usefully writes on coincidence as not being a phenomenon that can be discarded as unimportant for understanding how the world works:

> Coincidence, entirely proper to Virilio's age of the accident, is no more insubstantial than the vision that guides or the knowledge that must be analyzed and critiqued if it is to retain the status of knowledge – and which therefore must always be, like a vision, partial, temporary, and local.[25]

Google Earth is full of coincidence and chance – unexpected connections between people, places and time, and between the familiar and the unfamiliar. Our local knowledge is now connected to the local knowledge and experience of others in a vivid, previously inaccessible way. Google Earth also illustrates, via timelines and updates, how our local knowledge is always, as in Cubitt's

---

[24] Arthur Kroker and Marilouise Kroker, 'Introduction', in *Critical Digital Studies: A Reader*, Second Edition, eds Arthur Kroker and Marilouise Kroker (Toronto: University of Toronto Press, 2013), 7.
[25] Cubitt, *The Cinema Effect*, 326.

words quoted above, 'partial, temporary, and local'. With Google.com, we can use Google Earth to fly high and glimpse a fleeting sense of omniscience disguised as omnipresence. We can understand ourselves as good global citizens, or would-be spies, or any of the many roles we can take on as human avatars exploring a virtual globe of planet Earth. This social drama in which Google Earth is a part, is as much about scale as about any other aspect of the new technologies and their use.

Digital technologies have allowed humans to have an easily accessible view of themselves as part of a global society. As the most familiar of those strange objects called virtual globes, Google Earth plays a redressive role in the social drama of massive technological change and associated changes in the power relations that exist in many situations. It provides redress not as comfort or as a sign of willingness to change. It redresses the social drama of digital technologies by providing a way for us to account for our growing awareness and quest for understanding how 'things have changed' in our world. The advent of digital technology and our increasing codependence with our personal computers place us on a trajectory once again where the possibilities and problems incurred with the creation of new machinic companions also offer new insights into how we exist as humans.

References

Ackland, Robert, *Web Social Science. Concepts, Data and Tools for Social Scientists in the Digital Age*, Los Angeles, CA, London, New Delhi, Singapore and Washington DC: Sage, 2013.

Aday, Sean et al., *Blogs and Bullets: New Media in Contentious Politics*, Report from the United States Institute of Peace's Centers of Innovation for Science, Technology, and Peacebuilding, and Media, Conflict, and Peacebuilding (2010). http://www.usip.org/publications/blogs-and-bullets-new-media-in-contentious-politics (accessed 3 March 2014).

Akash, K.J., 'Missing Malaysian Airline Flight MH370: Don't Use Google Maps to Search for Plane, Says Google', *International Business Times* (11 March 2014), http://www.ibtimes.co.uk/missing-malaysia-airlines-flight-mh370-dont-use-google-maps-search-plane-says-google-1439744 (accessed 22 March 2014).

Andrejevic, Mark, 'Watching Back. Surveillance as Activism', in *Media and Social Justice*, eds Sue Curry Jansen et al., New York, NY: Palgrave Macmillan, 2011.

Arendt, Hannah, *Between Past and Future: Eight Exercises in Political Thought*, New York, NY: Viking, 1968.

Austin, T. and W. de Jong eds, *Rethinking Documentary: New Perspectives and Practices*, Maidenhead and New York, NY: McGraw Hill, 2008.

Australian Escape from Woomera Team, *Escape from Woomera* Computer Game.

Australian State of New South Wales, 'NSW Globe', http://globe.six.nsw.gov.au/ (accessed 6 April 2014).

Australian State of New South Wales (NSW) Department of Finance and Services, 'NSW Land and Property Information', 2014.

Australian State of Victoria Government, '2009 Victorian Bushfires Royal Commission Final Report', http://www.royalcommission.vic.gov.au/commission-reports/final-report (accessed 6 December 2013).

Baer, Ulrich, *Spectral Evidence: The Photography of Trauma*, Cambridge: MIT Press, 2002.

Bailey, J.E., S.J. Whitmeyer and D.G. De Paor, 'Introduction: The Application of Google Geo Tools to Geoscience Education and Research', in *Google Earth and Virtual Visualizations in Geoscience Education and Research*, eds S.J. Whitmeyer et al., Boulder, CO: Geological Society of America, Special Paper 492, 2012, vii–xix, doi: 10.1130/2012.2492(00).

Bassett, Caroline, 'Remote Sensing', in *Sensorium. Embodied Experience, Technology and Contemporary Art*, ed. Caroline A. Jones, Cambridge, MA and London: MIT Press, 2006.

Bateson, Gregory, *Steps to an Ecology of Mind. Collected Essays in Anthropology, Psychiatry, Evolution and Epistemology* (1972), Northvale, NJ and London: Jason Aronson Inc., 1987.

Baudrillard, Jean, *Simulacra and Simulation*, trans. Sheila Faria Glaser, Ann Arbor, MI: The University of Michigan Press, 2004.

Benjamin, Walter, *Illuminations*, ed. Hannah Arendt, trans. Harry Zorn, New York, NY: Shocken Books, 1969a.

———, 'Work of Art in the Age of Mechanical Reproduction' (1935), in *Illuminations*, ed. Hannah Arendt, trans. Harry Zorn, New York, NY: Shocken Books, 1969b, 217–251.

———, 'Berlin Childhood Around 1900', in *Walter Benjamin: Selected Writings* 3, 1935–1938, eds Howard Eiland and Michael W. Jennings, Cambridge, MA and London: The Belknap Press of Harvard University Press, 2002.

Berry, C., S. Kim and L. Spigel, eds, *Electronic Elsewheres. Media Technology and the Experience of Social Space*, Minneapolis, MN and London: University of Minnesota Press, 2010.

Betancourt, Michael, 'The Aura of the Digital', *Critical Digital Studies. A Reader*, eds Arthur Kroker and Marilouise Kroker, Second Edition, Toronto, Buffalo and London: University of Toronto Press, 2013.

Bookchin, Natalie and Blake Stimson, 'Out in Public: Natalie Bookchin in Conversation with Blake Stimson', in *Video Vortex Reader II: Moving Images Beyond YouTube*, eds Geert Lovink and Rachel Sommers Miles, Amsterdam Institute of Network Cultures, 2011, 306–317.

Boulos, Kamel et al., 'Web GIS in Practice X: A Microsoft Kinect Natural User Interface for Google Earth Navigation', *International Journal of Health Geographics* 10, no. 45 (2011). doi: 10.1186/1476-072X–10–45.

Brand, Stewart, 'Why Haven't We Seen the Whole Earth?', in *The Sixties: The Decade Remembered Now, by the People Who Lived It Then*, ed. Linda Rosen Obst, New York, NY: Rollingstone Press, 1977.

Brecht, Bertold, *Brecht on Theatre. The Development of an Aesthetic.* ed. and trans. J. Willett, London: Eyre Methuen, 1978.

Brighenti, Andrea Mubi, 'Artveillance: At the Crossroads of Art and Surveillance', *Surveillance and Society* 7, no. 2 (2010): 175–186. ISSN 1477-7487.

Butler, Judith, *Frames of War: When Is Life Grievable?* London and New York, NY: Verso, 2010.

Cavell Mertz and Assoc. Inc., FCCInfo.com http://www.fccinfo.com/cmdpro. php?sz=L&wd=1440

Chadwick, Andrew and Philip N. Howard, eds, *The Routledge Handbook of Internet Politics*, London and New York, NY: Routledge, 2009.

Chotaliya, Brinda M. and Sarang Masani, 'Remote Sensing: Essentials and Applications', *International Journal of Engineering Trends and Technology* 4, no. 8 (2013): 3460–3467. ISSN: 2231–5381.

Clarke, Philippa et al., 'Using Google Earth to Conduct a Neighborhood Audit: Reliability of a Virtual Audit Instrument', *Health and Place* 16 (2010): 1224–1229. doi: 10.1016/j.healthplace.2010.08.007.

Commonwealth Department of the Environment, Australia, http://www.environment.gov.au/aggregation/topics (accessed 9 January 2014).

Crampton, Jeremy W., 'Cartography: Maps 2.0', *Progress in Human Geography* 33, no. 2 (2009): 91–100.

Crosby, Chris, 'LiDAR Beginning to Appear in Google Maps Terrain Layer', in *OpenTopography Community Blog*, 30 July 2010. http://www.opentopography.org/index.php/blog/detail/lidar_beginning_to_appear_in_google_maps_terrain_layer (accessed 17 February 2014).

Crowder, David A., *Google Earth for Dummies*, Hoboken, NJ: Wiley Publishing, Inc., 2007.

Crutcher, Michael and Matthew Zook, 'Placemarks and Waterlines: Racialized Cyberscapes in Post-Katrina Google Earth', *Geoforum* 40 (2009): 523–534. doi: 10.1016/j.geoforum.2009.01.003.

Cubitt, Sean, *Digital Aesthetics*, London: Sage Publications, 1998.

———, *The Cinema Effect*, Cambridge, MA and London: MIT Press, 2005.

Cudd, Anne E. and Robin O. Andreasen, eds, *Feminist Theory: A Philosophical Anthology*, Malden, MA and Oxford: Blackwell Publishing, 2005.

Curry Jansen, Sue et al. eds, *Media and Social Justice*, New York, NY: Palgrave Macmillan, 2011.

Deleuze, Gilles and Félix Guattari, *A Thousand Plateaus. Capitalism and Schizophrenia*, trans. Brian Massumi, Minneapolis, MN and London: University of Minnesota Press, 1987.

de Zwort, Melissa, 'Cosplay, Creativity and Immaterial Labours of Love', in Amateur Media. Social, Cultural and Legal Perspectives, eds Dan Hunter et al., New York, NY: Routledge, 2013, 170–177.

DigitalGlobe.com

Dominguez, Ricardo, 'Electronic Civil Disobedience: Inventing the Future of Online Agitprop Theater', *PMLA* 124, no. 5 (2009): 1806–1812.

Dovey, Jon, 'Simulating the Public Sphere', in *Rethinking Documentary: New Perspectives and Practices*, eds T. Austin and W. de Jong, Maidenhead, McGraw Hill, 2008, 232–245.

Dovey, Jon and Helen W. Kennedy, *Game Cultures: Computer Games as New Media*, Maidenhead: Open University Press, 2006.

Dovey, Jon and Rose, Mandy, 'We're Happy and We Know It: Documentary, Data, Montage', *Studies in Documentary Film* 6, no. 2 (2012): 160–173.

Eiland, Howard and Michael W. Jennings, eds, *Walter Benjamin: Selected Writings* 3, 1935–1938, Cambridge, MA and London: The Belknap Press of Harvard University Press, 2002.

Erikson, J., 'The Face and the Possibility of an Ethics of Performance', *Journal of Dramatic Theory and Criticism* XIII, no. 2 (Spring 1999): 5–22.

ESA Earth Net, Online https://Earth.esa.int/web/guest/missions/3rd-party-missions/current-missions/ikonos-2 (accessed 5 November 2013).

ESRI www.esri.com

Facebook.com

Farman, Jason, 'Mapping the Digital Empire: Google Earth and the Process of Postmodern Cartography', *New Media and Society* 12, no. 6 (2010): 869–888. doi: 10.1177/1461444809350900.

Farman, Jason, *Mobile Interface Theory. Embodied Space and Locative Media*, New York, NY and London: Routledge, 2012.

Frankel, Felice, *Science* 280 (1998), 1698–1700, quoted in the Preface of The Research and Technology Organization (RTO) of NATO RTO Technical Report 30, 2001 © RTO/NATO ISBN 92-837-1066-5, 'Visualisation of Massive Military Datasets: Human Factors, Applications, and Technologies'. This Technical Report represents the Final Report of IST-013/RTG-002 submitted by the members of IST-013/RTG-002 for the RTO Information Systems Technology Panel (IST) M.M. Taylor (Canada), eds J.G. Hollands et al.

Frankel Paul, Ellen et al., eds, *The Communitarian Challenge to Liberalism*, Cambridge: Cambridge University Press, 1996.

Friedberg, Anne, *The Virtual Window. From Alberti to Microsoft*, Cambridge, MA and London: MIT Press, 2006.

Friedman, Ted, 'Civilization and Its Discontents: Simulation, Subjectivity, and Space', in *On a Silver Platter: CD-ROMs and the Promises of a New Technology*, ed. Greg Smith, New York, NY: New York University Press, 1999, 132–150.

Gadamer, Hans-Georg, 'The Historicity of Understanding as Hermeneutic Principle', in *Heidegger and Modern Philosophy*, ed. Michael Murray, New Haven, CT and London: Yale University Press, 1978, 161–183.

Gadamer, Hans-Georg, 'Text and Interpretation', in *Hermeneutics and Modern Philosophy*, ed. Brice R. Wachterhauser, trans. Dennis J. Schmidt, Albany, NY: State University of New York Press, 1986, 377–396.

Galloway, Alexander R., *The Interface Effect*, Cambridge and Malden, MA: Polity, 2012.

Gibson, William, *Neuromancer*, New York, NY: Ace Books, 1984.

Gilligan, Carol, *In a Different Voice. Psychological Theory and Women's Development* (1982), Cambridge, MA: Harvard University Press, 1993.

Goodchild, Michael F., 'Citizens as Sensors: The World of Volunteered Geography', *GeoJournal* 69 (2007): 211–221. doi: 10.1007/s10708-007-9111-y.

Goode, Luke, 'Social News, Citizen Journalism and Democracy', *New Media and Society* 11 (2009): 1287–1305. doi: 10.1177/1461444809341393.

Google.com

Google Earth Outreach Executive Interview (14 February 2014).

Graphic Design Forum, 'An explanation of Raster vs Vector', http://www.graphicdesignforum.com/forum/forum/graphic-design/resources/89-an-explanation-of-raster-vs-vector (accessed 8 August 2014).

Gregory, Sam, 'Transnational Storytelling: Human Rights, WITNESS, and Video Advocacy', *American Anthropologist* 108, no. 1 (2006): 195–204.

Guattari, Félix, *Chaosmosis. An Ethico-aesthetic Paradigm*, trans. Paul Bains and Julian Pefanis, Sydney: Power Publications, 1995.

Gunning, Tom, 'The Cinema of Attractions: Early Film, Its Spectator and the Avant-Garde', *Wide Angle* 8, nos. 3 and 4 (1986): 63–70.

Gurevitch, Leon, 'The Digital Globe as Climatic Coming Attraction: From Theatrical Release to Theatre of War', *Canadian Journal of Communication* 38 (2013): 333–356.

Habermas, Jurgen, *Theory and Practice*, trans. J. Viertel, London: Heinemann, 1973.

Hand, Seán ed., *The Levinas Reader*, Oxford and Cambridge, MA: Basil Blackwell, 1989.

Hansen, Mark B.N., *New Philosophy for New Media*, Cambridge, MA and London: MIT Press, 2006.

Haraway, Donna J., *Simians, Cyborgs and Women. The Reinvention of Nature*, London: Free Association Books, 1991.

Haraway, Donna J., 'A Cyborg Manifesto: Science Technology and Socialist Feminism in the Late 20th Century', in *The International Handbook of Virtual Learning Environments*, eds J. Weiss et al. Dordrecht: Springer, 2006, 117–158.

Haraway, Donna J., 'A Game of Cat's Cradle: Science Studies, Feminist Theory, Cultural Studies', in *Critical Digital Studies: A Reader*, Second Edition, eds Arthur Kroker and Marilouise Kroker, Toronto, Buffalo and London: University of Toronto Press, 2013, 59–69.

Hawkins, Virgil, 'Creating a Groundswell or Getting on the Bandwagon? Celebrities, Media and Distant Conflict', in *Transnational Celebrity Activism in Global Politics*, eds L. Tsaliki, C. A. Franonikolopoulos and A. Huliaras, Bristol and Chicago: Intellect and Chicago University Press, 2011, 85–104.

Hayles, N. Kathryne, *How We Became Posthuman*, Chicago and London: University of Chicago Press, 1999.

Hesford, W.S., 'Documenting Violations: Rhetorical Witnessing and the Spectacle of Distant Suffering 1', *Biography* 27, no. 1 (2004): 104–144.

Hinderliter, Beth et al., 'Introduction', in *Communities of Sense. Rethinking Aesthetics and Politics*, eds Beth Hinderliter et al., Durham and London: Duke University Press, 2009, 1–28.

——, eds, *Communities of Sense. Rethinking Aesthetics and Politics*, Durham and London: Duke University Press, 2009.

Hollinger, Andrew, 'United States Holocaust Memorial Museum Crisis in Darfur' on the Google Earth Outreach web page http://www.google.com.au/earth/outreach/stories/darfur.html (accessed 12 February 2014).

Holquist, Michael, *Dialogism: Bakhtin and His World*, London: Routledge, 1990.

Hui Kyong Chun, Wendy and Thomas Keenan, eds, *New Media, Old Media: A History and Theory Reader*, New York, NY and Abingdon: Routledge, 2006.

Hunter, Dan et al., eds, *Amateur Media. Social, Cultural and Legal Perspectives*, New York, NY: Routledge, 2013.

James Smith, Debbie, 'Big-Eyed, Wide-Eyed, Sad-Eyed Children: Constructing the Humanitarian Space in Social Justice Documentaries', *Studies in Documentary Film* 3, no. 2 (2009): 159–175. doi: 10.1 386/sdf.3.2.159/1.

Jameson, Fredrik, *The Prison-House of Language: A Critical Account of Structuralism and Russian Formalism*, Princeton: Princeton University Press, 1972.

Jenkins, Henry, *Convergence Culture: Where Old and New Media Collide*, New York: New York University Press, 2006.

Jones, Caroline A., ed., *Sensorium: Embodied Experience, Technology and Contemporary Art*, Cambridge and London: MIT Press, 2006.

Kember, Sarah and Joanna Zylinska, *Life After New Media. Mediation as a Vital Process*, Cambridge, MA and London: MIT Press, 2012.

Kingsbury, Paul and John Paul Jones III, 'Walter Benjamin's Dionysian Adventures on Google Earth,' *Geoforum* 40, no. 4 (2009): 502–513. ISSN 0016–7185.

Kozel, Susan, *Closer, Performance, Technology, Phenomenology*, Cambridge, MA and London: MIT Press, 2007.

Kroker, Arthur and Marilouise Kroker, 'Introduction', in *Critical Digital Studies: A Reader*, Second Edition, eds Arthur Kroker and Marilouise Kroker, Toronto, Buffalo and London: University of Toronto Press, 2013, 3–37.

Kuhn, Annette and Guy Westwell, eds, *A Dictionary of Film Studies*, Oxford: Oxford University Press, 2012, online version 2014. eISBN 9780191744426.

Latour, Bruno, *Reassembling the Social. An Introduction to Actor-Network-Theory*, Oxford: Oxford University Press, 2007.

Lefebvre, Henri, *The Production of Space*, Oxford: Blackwell Publishing, 1991.

Lenoir, Tim, 'Affect as Interface: Confronting the "Digital Facial Image"', in *New Philosophy for New Media*, ed. Mark B.N. Hansen, Cambridge, MA and London: MIT Press, 2006.

Levinas, Emmanuel, 'Ethics as First Philosophy', in *The Levinas Reader*, ed. Seán Hand, Oxford Cambridge, MA: Basil Blackwell, 1989.

Lovink, Geert and Rachel Sommers Miles, eds, *Video Vortex Reader II: Moving Images Beyond YouTube*, Amsterdam Institute of Network Cultures, 2011.

Luccio, Matteo, 'Google Earth Builder. Productizing Server Farms for Storing and Processing Geospatial Data', *Imaging Notes* 27, no. 2 (2012): 1–5. http://www.imagingnotes.com/go/article_freeJ.php?mp_id=303 (accessed 6 April 2014).

Manovich, Lev, *The Language of New Media* Cambridge, MA and London: MIT Press, 2000.

——, 'The Archaeology of Windows and Spatial Montage' (September 2002). http://www.manovich.net/DOCS/windows_montage.doc (accessed 15 January 2013).

MapAction, 'Google Earth and its potential in the humanitarian sector: a briefing paper', 2008. http://www.mapaction.org

Marmor, Kathy, 'Bird Watching: An Introduction to Amateur Satellite Spotting', *Leonardo* 41, no. 4 (2008): 317–323.

Massumi, Brian, 'Notes on the Translation and Acknowledgments. Pleasures of Philosophy', in *A Thousand Plateaus. Capitalism and Schizophrenia*, eds Gilles Deleuze and Félix Guattari, Minneapolis, MN and London: University of Minnesota Press, 1987.

——, 'The Autonomy of Affect', 1995. http://www.brianmassumi.com/textes/ Autonomy%20of%20Affect.PDF (accessed 30 January 2014).

——, 'The Bleed: Where Body Meets Image', in *Rethinking Borders*, ed. J.C. Welchman, London: Macmillan Press, 1996.

——, *Semblance and Event. Activist Philosophy and the Occurrent Arts*, Cambridge, MA: MIT Press, 2011.

McCaughey, Martha and Michael D. Ayers, eds, *Cyber-activism: Online Activism in Theory and Practice*, New York, NY: Routledge, 2003.

McClendon, Brian, 'Acceptance Speech on being awarded the UNEP Champions of the Earth Award in New York' 16 September 2013. Accessed via vimeo, http://vimeo.com/79463572 (accessed 17 February 2014).

McLuhan, Marshall, *Understanding Media. The Extensions of Man*, New York, NY: McGraw Hill, 1964.

——, 'A Dialogue: Gerald E. Stearn and Marshall McLuhan', in *McLuhan Hot and Cool*, ed. Gerald Emanuel Stearn, Harmondsworth and Ringwood Victoria: Penguin Books, 1968.

——, 'Quentin Fiore and Co-ordinated by Jerome Agel', in *The Medium Is the Massage. An Inventory of Effects*, Berkeley, CA: Bantam Books/Random House, 1967, Ginko Press, 2000.

McPherson, Tara, 'Reload: Liveness, Mobility, and the Web', in *New Media, Old Media: A History and Theory Reader*, eds Wendy Hui Kyong Chun and Thomas Keenan, New York, NY and Abingdon: Routledge, 2006, 199–208.

McVeigh, Tracy, 'Iranian Fugitive: Identity Mix-up with Shot Neda Wrecked my Life', *The Observer* (14 October 2012) http://www.theguardian.com/world/2012/oct/14/iran-neda-soltani-id-mix-up (accessed 3 March 2014).

Merleau-Ponty, Maurice, *Phenomenology of Perception* (1945), trans. Colin Smith, London and New York, NY: Routledge, 2004.

Merriam-Webster Online Dictionary.

Moore, Rebecca, 'Raising Global Awareness with Google Earth. What Do You Do After Flying to Your Home?' *Imaging Notes* 22, no. 2 (2007): 1–7.

——, 'Introducing Google Earth Outreach', Google Official Blog, 26 June 2007. http://googleblog.blogspot.com.au/2007/06/introducing-google-earth-outreach.html (accessed 31 March 2014).

——, 'Google Earth Outreach: Seeing Is Believing', in *Ngarluma Ngurra: Aboriginal Culture on the map* Catalogue, 2012, 19–20. http://issuu.com/form-wa/docs/ngarluma_ngurra_catalogue_4ec7aebf3b297f, 19–20. (accessed 31 March 2014).

Morley, David, 'Domesticating Dislocation in a World of "New" Technology', in *Electronic Elsewheres. Media Technology and the Experience of Social Space*, eds

C. Berry, S. Kim and L. Spigel, Public Worlds 17, Minneapolis, MN and London: University of Minnesota Press, 2010, 3–16.

Mouffe, Chantal, *On the Political*, London: Routledge, 2005.

Munster, Anna, *Materializing New Media. Embodiment in Information Aesthetics*, Hanover, NH: University Press of New England, 2006.

——, *An Aesthesia of Networks. Conjunctive Experience in Art and Technology*, Cambridge, MA: MIT Press, 2013.

Myers, Adrian, 'Camp Delta, Google Earth and the Ethics of Remote Sensing in Archaeology', *World Archeology* 42, no. 3 (2010): 455–467. doi: 10.1080/00438243.2010.498640.

Nash, Kate, Craig Hight and Catherine Summerhayes, eds, *New Documentary Ecologies. Emerging Platforms, Practices and Discourses*, Houndsmill: Palgrave Macmillan, 2014.

Natsios, Andrew S., *Sudan, South Sudan, and Darfur. What Everyone Needs to Know*, Oxford and New York, NY: Oxford University Press, 2012.

Ngarluma Ngurra: Aboriginal Culture on the map, http://issuu.com/form-wa/docs/ngarluma_ngurra_catalogue_4ec7aebf3b297f (accessed 31 March 2014).

Ngarluma Ngurra Google Earth Tour, http://www.form.net.au/project/ngarluma-ngurra/ (accessed 31 March 2014).

Norris, Pat, *Watching Earth from Space: How Surveillance Helps Us – and Harms Us*, Dordrecht: Springer, 2010.

Nussbaum, Martha, 'Compassion: The Basic Social Emotion', in *The Communitarian Challenge to Liberalism*, eds Ellen Frankel et al., Cambridge, New York, NY and Melbourne: Cambridge University Press, 1996, 27–58.

Nyerges, Timothy L., Helen Couclells and Robert McMaster eds, *The Sage Handbook of GIS and Society*, London: Sage Publications, 2011, doi: http://dx.doi.org.virtual.anu.edu.au/10.4135/9781446201046.

Official Google Blog, http://googleblog.blogspot.com.au/

O'Neil, Mathieu, *Cyberchiefs. Autonomy and authority in Online Tribes*, London and New York, NY: Pluto Press, 2009.

Online Description of 'RTO-TR-030 – Visualisation of Massive Military Datasets: Human Factors, Applications, and Technologies', *Scientific Publications of NATO Research & Technology Organisation*, Tuesday 1 March 2001, http://nato-pubs.ekt.gr/NATORTO/handle/123456789/4541 (accessed 6 April 2014). ISBN 92-837-1066-5.

OpenTopography.org

O'Reilly, Tim, 'What Is Web 2.0: Design Patterns and Business Models for the Next Generation of Software', *Communications and Strategies* 65, no. 1 (2007): 17–37.

Oxford Reference Online. Oxford University Press.

Papacharissi, Zizi, 'The Virtual Sphere 2.0. The Internet, the Public Sphere, and Beyond', in *The Routledge Handbook of Internet Politics*, eds Andrew Chadwick and Philip N. Howard, London and New York, NY: Routledge, 2009.

Parks, Lisa, *Cultures (Satellites and the Televisual) in Orbit*, Durham and London: Duke University Press, 2005, 230–245.

———, 'Digging into Google Earth: An Analysis of "Crisis in Darfur"', *Geoforum* 40, no. 4 (2009): 535–545.

———, 'Orbital Performers and Satellite Translators: Media Art in the Age of Ionospheric Exchange', *Quarterly Review of Film and Video* 24, no. 3 (2010): 207–216.

———, 'When Satellites Fall: On the Trails of Cosmos 954 and USA 193', in *Down to Earth*, eds Lisa Parks and James Schwoch, New Brunswick, NJ and London: Rutgers University Press, 2012, 221–237.

———, 'Earth Observation and Signal Territories: Studying U.S. Broadcast Infrastructure through Historical Network Maps, Google Earth, and Fieldwork', *Canadian Journal of Communication* 38 (2013): 285–307.

Parks, Lisa and James Schwoch, *Down to Earth*, New Brunswick, NJ and London: Rutgers University Press, 2012.

Pearce, G. and McLaughlin, C. eds, *Truth or Dare. Art and Documentary*, Bristol and Chicago: Intellect, 2007.

Raley, Rita, *Tactical Media*, Minneapolis, MN and London: University of Minnesota Press, 2009.

Rancière, Jacques, 'The Distribution of the Sensible: Politics and Aesthetics', *The Politics of Aesthetics*, trans. Gabriel Rockhill, London and New York, NY: Continuum, 2004.

———, *The Emancipated Spectator*, trans. Gregory Elliott, London and New York, NY: Verso, 2009.

Renov, Michael, in 'Collaborations and Technologies. Stella Bruzzi (interlocutor), Gideon Koppel, Jane and Louise Wilson in Conversation', in *Truth or Dare. Art and Documentary*, eds G. Pearce and C. McLaughlin, Bristol: Intellect, 2007, 65–80.

Research and Technology Organization (RTO) of NATO RTO, Technical Report 30, 2001 © RTO/NATO ISBN 92-837-1066-5, 'Visualisation of Massive Military Datasets: Human Factors, Applications, and Technologies'. This Technical Report represents the Final Report of IST-013/RTG-002 submitted by the members of IST-013/RTG-002 for the RTO Information Systems Technology Panel (IST) M.M. Taylor (Canada), eds J.G. Hollands et al. (2010).

Richards, John A., *Remote Sensing Digital Image Analysis. An Introduction*, Fifth edition, Berlin, Heidelberg: Springer Publications, 2013, doi: 10.1007/978–3–642–30062–2.

Russill, Chris, 'Earth-Observing Media', *Canadian Journal of Communication* 38, no.3 (2013): 277–284. ISSN 1499–6642.

Satellite Industry Association (SIA), 'State of the Satellite Industry Report, June 2013'.

Satellite Sentinel Project, http://www.satsentinel.org/our-story (accessed 25 May 2013).

Schachtman, Noah, 'What Did Google Earth Spot in the Chinese Desert? Even an ex-CIA Analyst Isn't Sure', *Wired* (1 September 2013), http://www.wired.com/dangerroom/2013/01/google-earth-china-hunh/#slideid-130941 (accessed 18 December 2013).

Schechner, Richard, Chapter 4, 'Play', in *Performance Studies: An Introduction*, Third edition, London and New York, NY: Routledge, 2013.

Smith, Colin, 'Photoshop Layers, 101', http://www.photoshopcafe.com/tutorials/layers/intro.htm (accessed 8 August 2014).

Smith, Greg, ed., *On a Silver Platter: CD-ROMs and the Promises of a New Technology*, New York, NY: New York University Press, 1999.

Sobchack, Vivian, *The Address of the Eye: A Phenomenology of Film Experience*, Princeton: Princeton University Press, 1992.

Sorensen, Andrew, ed., *Australian Computer Music Conference 2009*, Melbourne: Australasian Computer Music Association, 2009.

Stahl, Roger, 'Becoming Bombs: 3D Animated Satellite Imagery and the Weaponization of the Civic Eye', *MediaTropes* 11, no. 2 (2010): 113–134. ISSN 1913–6005.

Stearn, Gerald E., ed., *McLuhan Hot and Cool*, Harmondsworth: Penguin Books, 1968.

Steel, Jonathon, 'Violence Flares in Darfur's Kalma Refugee Camp as a New Cycle of Persecution Begins', *The Guardian* (27 October 2007), http://www.guardian.co.uk/world/2007/oct/27/sudan.international (accessed 24 January 2013).

Stensgaard, Anna-Sofie et al., 'Virtual Globes and Geospatial Health: The Potential of New Tools in the Management and Control of Vector-borne Diseases', *Geospatial Health* 3, no. 2 (2009): 127–141. www.GnosisGIS.org

Summerhayes, Catherine, 'Google Earth and the Business of Compassion', *Global Media Journal: Australian* 4, no. 2 (2010): 1–14. ISSN 1835–2340.

———, 'Embodied Space in Google Earth: *CRISIS IN DARFUR*', *Media Tropes* 3, no. 1 (2011): 113–134.

———, 'Web-Weaving: The Affective Movement of Documentary Imaging', in *New Documentary Ecologies. Emerging Platforms, Practices and Discourses*, eds Kate Nash, Craig Hight and Catherine Summerhayes, Houndsmill and New York, NY: Palgrave Macmillan, 2014, 83–102.

Sundvall, Erik et al., 'Graphical Overview and Navigation of Electronic Health Records in a Prototyping Environment Using Google Earth and openEHR Archetypes', in Proceedings of the 12th World Congress on Health (Medical) Informatics MEDINFO 2007, eds K. Kuhn et al., IOS Press, 2007, 1043–1047. http://www.iospress.nl/bookserie/studies-in-health-technology-and-informatics/

Swift, Ben, H. Gardner and A. Ridell, 'Distributed Performance in Live Coding stuff', in *Australian Computer Music Conference 2009*, ed. Andrew Sorensen Melbourne: Australasian Computer Music Association, 2009, 1–16.

Taylor, M.M., ed. (Canada) Research and Technology Organisation of NATO Technical Report 30, 'Visualisation of Massive Military Datasets: Human Factors, Applications, and Technologies', © RTO/NATO (2001).

The Shorenstein Center on Media, Politics and Public Policy, Harvard Kennedy School, 'Journalist's Resource', http://journalistsresource.org

The Tauri Group for the Satellite Industry Association (SIA), 'State of the Satellite Industry Report, June 2013'.

Torchin, Leshu, *Creating the Witness. Documenting Genocide on Film, Video, and the Internet*, Minneapolis, MN and London: University of Minnesota Press, 2012.

Tronto, Joan, *Moral Boundaries*, New York, NY: Routledge, 1993.

———, 'An Ethic of Care' (1993) reprinted in *Feminist Theory: A Philosophical Anthology*, eds Anne E. Cudd and Robin O. Andreasen, Malden, MA and Oxford: Blackwell Publishing, 2005, 251–263.

Tsaliki, L., C.A. Franonikolopoulos and A. Huliaras, eds, *Transnational Celebrity Activism in Global Politics*, Bristol: Intellect and Chicago University Press, 2011.

Turkle, Sherry, 'Tethering', in *Sensorium: Embodied Experience, Technology and Contemporary Art*, ed. Caroline A. Jones, Cambridge and London: MIT Press, 2006, 220–226.

———, *Alone Together. Why We Expect More from Each Other and Less from Technology*, New York, NY: Basic Books, 2011.

Turner, Victor, *The Anthropology of Performance*, New York, NY: PAJ Publications, 1986.

United Nations Environmental Program (UNEP) News Centre, 'Brian McClendon, co-founder and VP of Google Earth Awarded Top UN Environment Prize for Mapping New Conservation Paths and Creating Livelihood Opportunities through the Green Economy', http://www.unep.org/NewsCentre/default.aspx? DocumentID=2726&ArticleID=9621 (accessed 31 March 2014).

United States Holocaust Memorial Museum (USHMM), http:www.ushmm.org/

Vaidhyananthan, Siva, *The Googlization of Everything (and Why We Should Worry)*, Berkeley, CA: University of California Press, 2011.

van Dijck, José, *The Culture of Connectivity. A Critical History of Social Media*, Oxford and New York, NY: Oxford University Press, 2013.

2009 Victorian Bushfires Royal Commission Final Report, http://www.royalcommission. vic.gov.au/commission-reports/final-report (accessed 6 December 2013).

Virilio, Paul, *War and Cinema. The Logistics of Perception*, trans. Patrick Camiller, London and New York, NY: Verso, 1989.

———, *The Original Accident*, trans. Julie Rose, Cambridge and Malden, MA: Polity Press, 2005.

———, *City of Panic*, trans. Julie Rose, Oxford and New York, NY: Berg, 2007.

Vork Robert, 'Things That No One Can Say: The Unspeakable Act in Artaud's *Les Cenci*', *Modern Drama* 56, no. 3 (2013): 306–326.

Weiss, J. et al., eds, *The International Handbook of Virtual Learning Environments*, Dordrecht: Springer, 2006.

Welchman, J.C., ed., *Rethinking Borders*, London: Macmillan Press, 1996.

Whitmeyer, S.J. et al., eds, *Google Earth and Virtual Visualizations in Geoscience Education and Research*, Geological Society of America Special Paper 492 (2012). doi: 10.1130/2012.2492(00).

Willemen, Paul and Meaghan Morris, *Looks and Frictions. Essays in Cultural Studies and Film Theory*, London: BFI Publishing, 1993.

WITNESS.org

Wood, Jo et al., 'Interactive Visual Exploration of a Large Spatio-Teimporal Dataset: Reflections on a Geovisualization Mashup', *IEEE Transactions on Visualization and Computer Graphics* 13, no. 6 (2007): 1176–1183. doi: 10.1109/TVCG.2007.70570.

Yamagishi, Yasuko et al., 'Visualization of Geoscience Data on Google Earth: Development of a Data Converter System for Seismic Tomographic Models', *Computers and Geosciences* 36 (2010): 373–382. ISSN 0098-3004.

Zook, Matthew and Mark Graham, 'The Creative Reconstruction of the Internet: Google and the Privatization of Cyberspace and DigiPlace', *Geoforum* 38, no. 6 (2007): 1322–1343.

Index

CPSIA information can be obtained
at www.ICGtesting.com
Printed in the USA
LVOW10s1844161117

556557LV00012B/1024/P